This book is a catalogue of the chromatograms presently available obtained by the high-speed separation technique. It presents an outline of the theory and technique of HSLC for those not familiar with the subject. It will be invaluable to prospective users of this technique in that it gives in graphical form the results obtained by experienced workers in the field. In this book the method has been used for about 150 analytical problems of widely ranging difficulty and covering many different chemical types. The chromatograms are presented in a standard format from which the reader can see at a glance what were the exact experimental conditions used.

Applications of High-speed
Liquid Chromatography

Applications of High-speed Liquid Chromatography

John N. Done and **John H. Knox**
Department of Chemistry, University of Edinburgh

and

Joël Loheac
Perkin-Elmer, Boix-Colombes, France

A Wiley—Interscience Publication

JOHN WILEY & SONS

LONDON · NEW YORK · SYDNEY · TORONTO

Library of Congress Cataloging in Publication Data:

Done, John N.
Applications of high-speed liquid chromatography.

"A Wiley–Interscience publication."

1. Liquid chromatography. I. Knox, John H., joint author. II. Loheac, Joel, joint author. III. Title.

QD79.C454D66   544'.924   74–16148
ISBN 0 471 21784 0

Printed in Great Britain
By Unwin Brothers Limited
The Gresham Press, Old Woking, Surrey, England
A member of Staples Printing Group

PREFACE

What has come to be known as modern liquid chromatography, high-speed, high-performance, or high-efficiency liquid chromatography (HSLC) first came into prominence at the Fifth International Symposium on Advances in Chromatography held in Las Vegas in 1969. Since then HSLC has expanded rapidly with a publication rate doubling each 18 or 20 months. Undoubtedly the technique is now beginning to have a major impact on chemical analysis, and will shortly take its place alongside gas chromatography (GC), mass spectrometry, NMR, and other spectroscopic techniques as one of the major methods for separation, quantitation and identification of chemical substances. HSLC is the perfect complement to gas chromatography for it can separate just those compounds which because of involatility, or instability, cannot be handled by GC. These include polymers, most ionic compounds, a great variety of complex molecules with moderate to high molecular weights, the wide range of unstable and reactive compounds familiar to organic chemists, and a great variety of naturally occurring substances important in medicine and biochemistry. Also amenable to HSLC are those compounds of high polarity which can only be submitted to GC after derivitization, for example steroids, sugars and compounds of low molecular weight containing strongly polar functional groups.

Since HSLC is relatively new to most analysts, the majority of prospective users are not yet sufficiently confident of its value in their particular field to be able to justify spending a large sum of money on a commercial instrument. They require assurance—often from an instrument company—that the problem they have in mind can be solved by HSLC. Then there are those who have only just heard of HSLC and wish to know more of its range, while on the other hand, many who already have some practical experience wish to extend the range of application of their instrument(s) and require data on problems to which HSLC has been successfully applied.

This book is aimed at these groups, and it is essentially a survey of applications of HSLC up to about the end of 1973. We have selected examples from the literature which seem chemically significant, and we have tried to include examples from all areas of chemistry where modern HSLC methods have been used, even when the results leave something to be desired in terms of speed and resolution. This is intentional for the poorer chromatograms illustrate those areas where separations are currently difficult and they present a challenge to users of HSLC to do better.

The book falls naturally into two parts. The first briefly outlines the main features of equipment and of the principles of modern liquid chromatography; the second presents about 150 selected chromatograms taken from the recent literature (1969–1973) along with some unpublished material obtained in Edinburgh. Inevitably with the rapid advance of HSLC some of the chromatograms presented will be rapidly superceded but the nature of the operating conditions required for each separation will change less. We therefore believe that the data presented will continue to be a useful guide for some years to come.

In producing this book we are greatly indebted to Mr Colin Baxter who has drawn all the figures and chromatograms, the latter having been traced from enlargements of the originals and finally reduced to the standard format. We are equally indebted to Mrs Janice Henderson who produced the first class copy-ready typescript used in the final production.

J.N. Done
J.H. Knox
J. Loheac

June 1974

CONTENTS

PART 1

**THE TECHNIQUE AND THEORY OF HIGH
SPEED LIQUID CHROMATOGRAPHY**

CHAPTER 1

HISTORY AND GENERAL DESCRIPTION

The invention of chromatography in its modern form is generally ascribed to Tswett (1) who around 1903 first showed how compounds could be separated by elution through a column of adsorbent. Separation, as Tswett recognised, arose because of the different affinities of substances for the adsorbent with which the column was packed. From 1903 until his death in 1919, Tswett developed and improved his methods but his work was premature and the potential of chromatography was unrecognised by chemists at large who considered it to be little different from single stage extraction. The method was revived in 1931 by Kuhn and Lederer (2) who, following Tswett almost exactly, used chromatography to separate plant pigments. There was, however, little advance in either the theory or practice of chromatography until 1941 when Martin and Synge (3) published their classic paper describing the invention of liquid-liquid partition chromatography in which the different rates of migration of substances arose from their different partition ratios between the flowing mobile phase and a stationary liquid phase held in the interstices of a porous support. In their case water supported on silica gel was the stationary phase, and chloroform the mobile phase.

In this paper Martin and Synge (3) revealed an extraordinary grasp of the fundamentals of chromatography. They developed the concept of the 'height equivalent to a theoretical plate' which has become universally accepted as the measure of the efficiency of a chromatographic column; they showed how the elution time of a solute was directly connected with the partition ratio of the substance between the two phases in the column; they outlined the advantages of using a gas as mobile phase in place of a liquid and so foreshadowed gas chromatography successfully demonstrated experimentally only in 1952 (4). They also stated that fast analysis by liquid chromatography would require the use of very small particles and high pressure differences across the length of the column. In this they indicated the essential conditions for high speed liquid chromatography which have only been realised experimentally within the last five years (5, 6, 7). While the culmination of high speed liquid chromatographic systems (HSLC systems) can be pinpointed to 1969 (5, 6), pioneering work had been carried out as long ago as 1960 by Hamilton (8). Unfortunately this work, in the rather specialised field of amino-acid analysis, had little impact on liquid chromatography in general, and the first papers which showed analyses by liquid chromatography comparable in speed to those of gas chromatography did not appear until 1968 or 1969.

Almost without exception HSLC systems were developed by physical and

3

analytical chemists who had previously been prominent in the field of gas chromatography. The staggering success of gas chromatography arose because of its versatility, its sensitivity, its ease of operation, and its capacity for routine and automatic application. Its advance was greatly assisted by technological developments in electronics and by the development of new materials, but also by the development of theoretical models which led to an excellent understanding of the basic chromatographic processes. By contrast the practical and theoretical development of liquid chromatography lagged far behind, Hamilton (8) being one of the few people who had tried to apply the theory of gas chromatography in an attempt to improve the method.

Practical experience with gas chromatography coupled with the theoretical ideas established by Martin and Synge (3) and elaborated by Giddings (9) showed that there should be no insuperable difficulties in developing systems for liquid chromatography which would possess the advanced features of gas chromatography. In the ideal HSLC system, eluent would be pumped into the column under high pressure and at a precisely controlled rate; injection of the sample into the column would be either by an injection valve or by a syringe/septum arrangement, the sample size would be small to optimise resolution; the column itself would be long and narrow and packed with a carefully prepared and well-sized material composed of particles at least ten times smaller than those conventionally used in gas chromatography; finally, the emergent solution would be passed to a sensitive post-column detector which would monitor the concentration of solutes, and would produce an electrical signal suitable for recording. The credit for the simultaneous combination of all these features in a single system goes jointly to Kirkland (5) and Huber (6), and the present state of the art of HSLC has recently been summarised in Kirkland's book (10).

The essential features of a modern HSLC system are shown in Figure 1. Solvent is degassed in a reservoir by heating, evacuating or submitting to ultrasonic vibration. It then passes via a fine mesh filter to a pump which is capable of delivering solvent against a pressure of up to 3000-5000 p s i (200-330 atm). The pressure downstream of the pump is measured by a Bourdon gauge or pressure transducer. The plumbing throughout is of stainless steel to minimise corrosion and to withstand high pressures. Injection is either by valve or syringe and septum, and is made as near to the column head as possible; indeed it should ideally be made into the column packing itself. The column is a heavy walled tube either of glass or stainless steel 2 to 5 mm in diameter and between 0.10 and 2 m in length. The inner surface of the tube must have a mirror finish. Columns are generally thermostatted either in air oven or by water pumped from a thermostatted bath.

The column packing consists of very fine particles which have generally been specially prepared. They must have a range of size within a factor of two but preferably 1.5 and, mean diameters between 5 and 50 µm. The most widely used diameter range is from 30-50 µm (around 400 mesh). Column packing materials may either be completely porous (for instance silica gel, alumina,

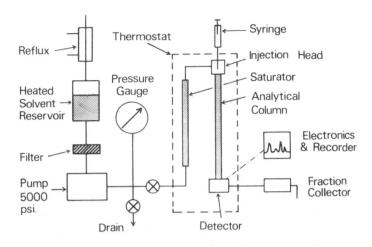

FIGURE 1. Outline of a high speed liquid chromatograph.

Chromosorb), or they may be superficially porous (for instance a thin porous layer
attached to a solid glass bead core). After passage through the column, the eluate
passes immediately via minimum volume connectors to the detector. Current
detectors have limits of detection between 10^{-6} and 10^{-9} mole fraction and the
total volume from the column exit to the end of the effective detector cell is
held preferably below 10 µl. The signal from the detector is electrical and is passed via suitable electronics to a recorder, integrator or other data handling system.

The record from a modern high speed liquid chromatograph is in the form
of an elution chromatogram, that is a record of detector response (proportional
to solute concentration) against time. All chromatograms shown in Part II of
this book are of this sort.

Generally a sophisticated HSLC system can be used for any of the many
forms of liquid chromatography. The chief of these are as follows:

1) Molecular exclusion chromatography (MEC) or gel permeation chromatography (GPC). Molecules are separated according to their size by passage down
a column packed with a microporous material, usually a porous glass or cross-
linked polystyrene for high pressure applications. Large molecules are excluded
from the pores of the packing while smaller molecules may permeate the pores
partially or totally depending upon their size. In exclusion chromatography the
largest molecules emerge from the column first and the smallest last. For this
form of chromatography it is important to avoid adsorption of the molecules on
to the surface of the porous material, so a polar eluent is normally used, and
often a silica based material is silanized by treatment with hexamethyldisilazane
vapour at 190°C.

2) Ion exchange chromatography (IEC). The column is packed either with

ion exchange resin beads or with surface layered ion exchange particles (pellicular materials). It is widely used in the analysis of amino acids, drugs, metabolites, pharmaceutical products, inorganic salts, and generally for compounds containing ionised or ionisable groups. The mobile phase in IEC is generally an aqueous buffer solution which may contain a small proportion of a water miscible organic solvent. Since the affinity of different compounds for an ion-exchange resin can vary enormously it is often convenient to change the composition of the eluent during elution of a sample. When this is done in a controlled way the technique is known as gradient elution or solvent programming. It is the equivalent in LC of temperature programming in GC.

3) Liquid-solid adsorption chromatography (LSAC). An adsorbent such as silica gel, alumina, or other material of large surface area is used as the partitioning material. LSAC is suitable for the separation of a wide range of organic materials which are soluble in non-polar or moderately polar organic solvents. As with IEC gradient elution can be used with excellent results.

4) Liquid-liquid partition chromatography (LLC). Here the stationary phase is a liquid supported on a solid. The supporting solid may be either a completely porous material such as a wide pore silica gel or porous glass, or a superficially porous (pellicular) material. LLC is applicable to a wide range of organic and inorganic solutes, and one can use either a polar stationary phase and a non-polar mobile phase, or the reverse. Since the two phases must remain distinct their mutual solubility should be small, and a precolumn must be included to saturate the mobile with stationary phase. Gradient elution is generally impossible with LLC because of the solubility problem.

5) Liquid-solid partition chromatography (LSPC). This is a relatively recent form of liquid chromatography in which the stationary phase is a polymer chemically bonded to the surface of the support (11, 12) which is normally silica. These materials are prepared by reacting the -OH groups of the silica surface with various reagents to give eventually →Si-C, →Si-N or →Si-O-C bonded materials of relatively high stability. The coating is essentially a solid layer, but has partition properties similar to those of a stationary liquid of similar composition. The great advantage of the bonded compared to the normal liquid stationary phases, is that they are insoluble in solvents and are thermally stable. They can thus be used most effectively in gradient elution systems.

CHAPTER 2

EQUIPMENT FOR HIGH SPEED LIQUID CHROMATOGRAPHY

HSLC systems differ chiefly in the methods used to generate high inlet pressures, to detect the eluted solvents, and to generate solvent programmes. Commercial instruments differ in their general layout and in the refinements which they include, for example, column thermostatting, pressure trip switches, solvent degassing systems, reservoir level gauges, fraction collectors and so on. Most instruments will be capable of performing all types of liquid chromatographic analysis with the exception of amino acid analysis since this, in its present form, demands a specific detection system of some complexity.

In this chapter we review the various types of equipment which are available for performing the main unit operations in HSLC.

1) Solvent preparation systems. As noted above it is important to degas and filter solvents before passage to the HSLC system proper. Solvents which are not degassed tend to produce air bubbles towards the column outlet and these interfere both with column performance and with the functioning of the detector. Very often a column is packed dry and it is necessary to remove the air from it before it can be used for chromatography. This is most simply done by passage of a solvent in which air is slightly soluble. For these reasons air free solvents are desirable. The solvents which give the most trouble in respect of dissolved gas are water and alcohols in which air is tolerably soluble. Degassing is best effected by reflux and evacuation, or evacuation coupled with ultrasonic vibration, but either reflux or evacuation alone will generally suffice. The degassing may be carried out independently of the chromatograph but many commercial instruments incorporate their own facilities. Filtration of the solvent before it enters the pump is desirable since many pumps depend upon the proper functioning of non return valves whose seating is critical and can be seriously impaired by small particles of dust. Metal or teflon filters of 5 or 10 micron porosity are satisfactory and are normally fitted immediately before the pump.

2) Pressurisation systems. Four types are in general use. In the simplest shown in Figure 2A the eluent is contained in a 10 m x 6 mm diameter coiled stainless steel tube and is driven into the column by the application of direct gas pressure. Solution of the gas in the eluent is restricted to the first few ml of the liquid, and this portion of the liquid is rejected each time the coil is refilled by arranging that filling takes place from the bottom of the coil. To fill the coil valves A and B are closed, valve C is opened to release the gas pressure, and valve D is opened, filling the reservoir by gravity. After filling the coil, valves C and D are closed, the system is pressurised by opening valve A and flow started by opening valve B. In commercial equipment the switching cycle is effected by a single control and safety features are included to prevent pressurisation of the

7

system when valves to the drain and reservoir are open. The maximum pressure is limited to that available in gas cylinders.

Indirect gas pressurisation by a pressure intensifier is safer. The principle is shown in Figure 2B. A relatively low gas pressure, say up to 150 psi (10 atm) acts on a piston of large area. This piston is rigidly connected to a piston of much smaller area (usually 1/20th or 1/40th) which bears directly on the liquid. The pressure of the gas is thus amplified 20 or 40 times. Automatic refill is generally fitted. With this type of system, the gas never contacts the liquid so it is impossible to fill the equipment with gas, an accident which can occur all too easily with direct gas pressure systems. Pressure intensifiers pressurise a relatively large volume of liquid, say 50 to 100 ml, and any gradient system (see section 7) must therefore operate on the high pressure side of the pump. Pressure intensifiers deliver liquid at a constant pressure not a constant flow rate. Since constant flow is often a requirement it is important to ensure that the column is thermostatted, to avoid viscosity changes and consequent flow change, and it is important to ensure that the column resistance does not change from other causes, for example, blocking of end fittings and frits, precipitation of polymers in the column, fragmentation of the packing, etc. Constant pressure systems do however, possess the important safety feature that they are not damaged by downstream blockage as flow simply stops. On the other hand leakage upstream of the main flow resistance will lead to the pump continually recycling until the supply of pressurisation gas is exhausted.

Motor drive syringes (Figure 2C) deliver a constant flow smoothly irrespective of back pressure, but of course, any downstream blockage can then cause a dramatic rise of pressure and some kind of safety valve is essential to prevent damage. Since syringe pumps like pressure intensifiers, hold a large volume of solvent, any gradient system must operate on the high pressure side of the pump. Syringe pumps tend to be expensive because of the very high quality seals and lead screws which are required.

The commonest pump is probably the reciprocating piston type with an adjustable stroke (figure 2D). Piston pumps deliver a controlled but pulsating flow of eluent. The piston may act either directly in the eluent, or it may act in an oil reservoir separated from the eluent by a diaphragm. In the latter the piston never contacts the eluent, so the seals may be optimised without having to consider corrosion problems or reactions of the eluent with the seals. However, the total liquid volume, oil plus eluent, in such pumps is larger than in the first type and compression of the total fluid gives a greater dependence of flow rate on back pressure. In all piston pumps the eluent is drawn through one non-return valve from the reservoir and pumped through a second non-return valve. These valves generally consist of small spherical balls of specially hardened steel or sapphire. Their seating is critical and depends on the cleanliness of the eluent which must be free from all particulate matter. Diaphragm piston pumps generally have very small volumes of eluent in front of the diaphragm and can often be used with gradient systems which operate on the low pressure side of the

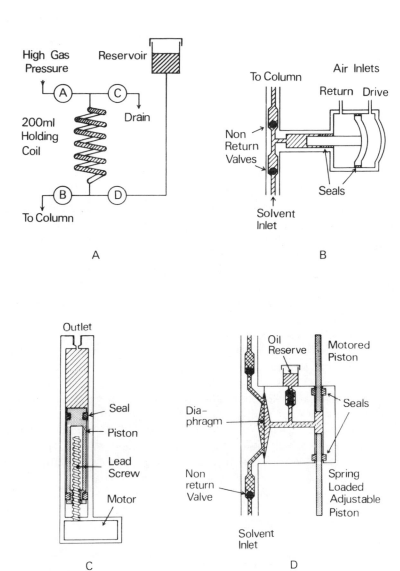

FIGURE 2. Pressurisation systems for high speed liquid chromatography. A - direct gas pressurisation; B - pressurisation with a pressure intensifier; C - pressurisation by motorised syringe; D - pressurisation by reciprocating pump.

pump. Their disadvantage is that they deliver a pulsating flow which must be smoothed. The simplest smoothing system is a Bourdon pressure gauge filled with air. Alternatively a long spiral may be used. Unfortunately nearly all pulse dampers introduce additional dead volume to the apparatus and make it more difficult to develop satisfactory low pressure gradient systems. A recent development is a form of pump in which two small volume reciprocating pumps operate alternately in such a way as to give an almost pulse free flow. The movements of the two pistons are so timed that there is a smooth transition from one cylinder to the other.

Pumps are amongst the most expensive components of HSLC systems, especially of home-made systems. The gas pressurisation system is the only one which can be constructed in the laboratory and is by far the cheapest. Unfortunately it is likely to be hazardous, and unless a well-designated commercial system with adequate safety features is purchased, direct gas pressurisation cannot be recommended for pressures in excess of 1000 psi Reciprocating piston pumps and pressure intensifiers cost around £600 while motor driven syringes and the pulseless piston systems cost between £1000 and £2000.

3) Detectors. Numerous detection systems have been reported but only three are widely used and generally satisfactory. These are the refractive index monitor (RI monitor) the ultra-violet photometer (UV photometer) and the wire transport system. Their relative merits have been discussed by Huber (13).

The RI monitor measures a general property of the eluate and in common with other general detectors it is relatively insensitive. The best available RI monitors will detect differences in RI between the unknown and reference stream of as little as 10^{-7} RI units. Unfortunately the total change in RI from pure eluent to pure solute is of the order of only 0.1 RI unit and thus the detection limit is around 1 part of solute in 10^6 of eluent. Even to achieve this level of sensitivity requires temperature stabilisation to about 10^{-3} °C. Heat exchangers are therefore necessary between the column and the detector. A sensitivity of 10^{-6} appears at first sight to be similar to that of a katharometer in gas chromatography, but it has to be remembered that a typical liquid eluent is roughly 500 times as dense as a gas. Thus the sensitivity on the basis of mass per unit volume of stationary phase is some 500 times less using a RI monitor in HSLC than it is using a moderately good katharometer in GC. Other general detectors based for example on comparison of dielectric constant, thermal conductivity, density, velocity of sound, suffer from similar limitations.

The UV monitor belongs to the general class of selective detectors in spite of its relatively wide applicability. It does, of course, require eluents which have no UV absorbance. For such solvents there is a large 'internal amplification' when a UV absorbing species passes through the detector cell, and a corresponding increase in sensitivity over that expected from a bulk property detector. Current UV monitors mostly operate with low pressure mercury lamps which deliver about 90% of their radiation at 254 nm, but variable wavelength photometers are now becoming more popular. Generally a double beam system is used to compensate for changes in light intensity. The light is monitored by matched photocells, photodiodes or photoresistors whose output is passed to logarithmic amplifiers and differenced. The difference signal is proportional to the absorbance and this, by Beer's law, to the concentration of absorbing species in the detector cell. Current models have noise levels in the region of 10^{-4} absorbance units, and so have detection levels of around 1 part in 10^9 for substances with molar extinction coefficients of 10^4 mol^{-1} l cm^{-1}. UV monitors are now the commonest detectors used in HSLC, but there is a great need for other detectors of similar sensitivity which will respond to compounds which do not absorb light in suitable regions of the spectrum. For both RI and UV monitors the effective cell volume including any connecting tubing from the column, must be small, and if possible the total should be below 10 μl. The internal dimensions of a typical UV monitor cell for example are 1 mm diameter and 10 mm long giving an effective cell volume of 7.5 μl. Such cells give negligible extra-column dispersion with unretained solutes eluted from 2 mm bore l m long columns packed with 30 μm particles. However when 5-10 μm diameter particles and 10 cm columns become common it may be necessary to reduce cell volumes to a few microlitres.

Other selective detectors are based upon measurement of fluorescence, colourimetry generally, polarographic reduction or oxidation, conductivity, and the occurrence of specific chemical reactions. Generally they may be expected to give sensitivities comparable with those of UV monitors.

In transport detectors the solute is physically separated from the eluent. The eluate is collected on a wire or band which passes first through an evaporator where the solvent, which must be volatile, is removed. The band then passes through a monitor which determines the quantity of solute remaining on it. In one form of the detector (14) the solute is burnt to carbon dioxide and water. The carbon dioxide is catalytically reduced to methane, and the methane monitored by a flame ionisation detector as used in gas chromatography. Although the principle of this detector is attractive, many problems have arisen in the design of satisfactory working instruments. Current models have sensitivities which are slightly better than those of RI monitors, but they still fall far short of UV monitors. Transport detection systems would appear to provide one of the best ultimate solutions to the problem of general detection in HSLC but there is much development work still to be done. Their greatest weakness at present stems from the small proportion of eluate picked by the transport mechanism. Since this is generally between 0.1 and 1% of the total, a factor of at least 100 in sensitivity is lost from this alone.

Current commercial detectors cost between £600 and £1200 and the reader is referred to Kirkland's 'Modern Practice of Liquid Chromatography' (10) for further details.

4) The column. The column and its contents are, of course, the core of any chromatograph. Columns may be made of heavy walled glass tubing or more generally of stainless steel. 2 mm bore, 8 mm outer diameter glass tubing can be

used up to pressures of 1500 psi (100 atm). For higher pressures stainless steel columns (preferably grade 316) must be used. Bright annealed stainless steel tubing as received from the manufacturers gives results which are inferior to those obtained with glass tubing, but the performance is much improved if the inside is polished. Columns are terminated at the downstream end by a tightly fitting teflon or stainless steel frit with a porosity of between 2 and 10 μm depending upon the size of the packing.

Column dimensions depend upon the particles size, the desired column efficiency (number of theoretical plates), and sometimes upon instrument configuration. Kirkland (5, 11, 15) in much of his work has used 30 μm particles in 2.1 mm bore columns 0.5 or 1 m in length. He found narrower columns difficult to pack and obtained poorer efficiencies with wider columns. However, with 8 and 11 mm bore columns, de Stefano and Beachell (16) obtained still higher efficiencies than with 2.1 mm columns. They showed that this was probably the result of the so-called 'infinite diameter effect'. When a sample is injected centrally into a sufficiently wide column lateral dispersion may be insufficient to carry the sample to the walls of the tube. In such circumstances wall effects, which spoil band profiles and reduce efficiency, are absent. The condition for this to be so was given by Knox and Parcher (17) as

$$Ld_p/d_c^2 < 0.4 \tag{1}$$

where L = column length, d_p = particles diameter, and d_c = column bore.

The first columns used by Kirkland, and still used for particles of 30 μm diameter and above, were sufficiently narrow that molecules made many traversals of the column before emerging. However, extremely high efficiencies are now being obtained with 5–20 μm particles in columns 5 mm bore and only 100 mm in length (18, 19). Such columns are effectively of infinite diameter and their excellent performance is due to this.

Since the volume of eluate containing an unretained or slightly retained band from a high efficiency column may be as little as 50 μl, extra-column volumes between the injector and detector must be kept to a minimum. Although it can be shown that 0.5 m of 0.15 mm (0.006") bore tubing has negligible effect on peak dispersion, quite small dead volumes within connectors and in the detector itself can have serious effects producing peak spreading and tailing. It is vital therefore to pay close attention to the geometry of all units. Some well tried fittings of good geometry are shown in Figure 3.

Samples are generally introduced into the column in dilute solution by microsyringe or injection valve. Samples are typically between 1 and 20 μl in volume and contain 0.1 to 10 μg of solute. Larger samples can be handled but peak shapes may then deteriorate because of overloading or excessive injection volumes.

The simplest injection method uses a syringe and captive septum as shown in Figure 3. The septum is tightly compressed by the retaining unit and only a small area is exposed and has to withstand the pressure differential. It is import-

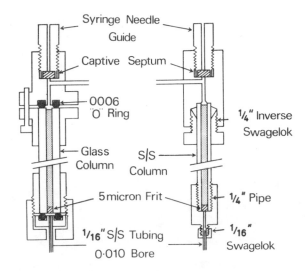

FIGURE 3.   Column inlet and outlet connections and injection block.

ant to arrange the diameter of the upper part of the unit to be just larger than
the syringe needle itself so that the needle passes through the same hole in the
septum each time a sample is injected. In this way septum life is maximised. The
lower hole should be chamfered so that the sharp end of the needle is not blunt-
ed by stiking the lower part of the injection unit. Septa can be used at least up to
1500 psi (100 atm). For higher pressure it is desirable to halt the flow before
injecting. Injection by sample valve is becoming increasingly popular and modern
valves will withstand pressure differences of 5000 psi (330 atm). A disadvantage
of injection valves is that they inject a constant sample volume and generally require a larger total volume of sample than is required to fill a microsyringe.

5) Column packing methods. For high performance liquid chromatography
particles of packing must be smaller than 50 μm diameter (275 mesh size), and
can be as small as 5 μm. Future trends will undoubtedly be towards the smaller
particles and ultimately 5 μm diameter particles may become the rule. Unless the
pressure capacity of HSLC equipment rises well beyond 5000 psi it is unlikely
that particles smaller than about 5 μm will be much used. In a typical column
between 2 and 5 mm bore there will be between 40 and 1000 particles across
the column. Because of the very large number of particles to the cross section it
is vital for good performance to pack the particles with the greatest possible uni-
formity. HSLC columns are in fact analogous to large scale preparative GC columns.

To produce high performance columns it is necessary to use well-graded
support or packing materials and a method of packing the column which avoids
fractionation. It is probably desirable to use particles whose size range does not
exceed a factor of about 1.5. Two packing methods have been developed which
avoid particle fractionation and give reproducibly high performance (11).

The best dry-packing method is the 'rotate, bounce and tap' method or RBT method. The column is mounted vertically and the packing slowly but continuously added while the column is being rotated at about 60 times per minute and bounced at about 100 times per minute. This can be carried out either by hand or preferably by machine as shown in Figure 4A. While adding the packing the column can beneficially be tapped with a light object at the level of the packing. The use of a vibrator must be avoided unless the particle size is very small. Then the intensity should be adjusted so that only about the top mm or so of packing is fluidized.

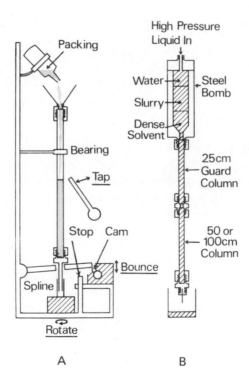

FIGURE 4. Methods of column packing. A) the 'Rotate, Bounce and Tap' method B) slurry packing.

The alternative to dry packing is to use a slurry technique. Ideally the supporting liquid should have the same density as the particles themselves as this prevents settling under gravity. For silica-based particles mixtures of iodo and bromomethanes can be used. Both the packing and supporting liquid must be carefully dried before making up the slurry. The balanced slurry is placed in a bomb attached to the top of the column (Figure 4B) which has already been filled with the supporting liquid or a denser solution. The slurry itself is carefully covered with more of the supporting liquid or a slightly less dense liquid and then with water. The slurry is then driven into the column under the maximum possible pressure, preferably 5000 psi or more. A hydraulic ram may be used. The essential for success is that the particles of the slurry impact on the growing bed with maximum force. Otherwise the packing will be unstable and will collapse when the column is used for chromatography, slurry packings generally being less densely packed than dry packings. Slurry packing methods are most effective with small non-spherical particles. An alternative to using the balanced slurry method is to employ a supporting solution which preserves a surface charge on the particles; silica, for example, may be supported in dil ammonia, alumina in dil acetic acid. The electrostatic repulsion of the particles maintains them in suspension and reduces fractionation if a dense slurry is used.

When a liquid stationary phase has to be placed on a packing this may be loaded as for gas chromatography. The packing is wetted with a solution of the stationary phase in a volatile solvent of sufficient volume to cover it completely. The solvent is then removed in a rotary evaporator. Alternatively the stationary phase may be deposited in situ (18). A 10% to 30% solution of stationary phase is made up in a suitable solvent and pumped into the column which has already been filled with the solvent. The solution is then displaced by a second solvent which will dissolve the first but will precipitate the stationary phase. An emulsion is thereby formed both in the flowing part of the mobile phase and in the interstices of the packing. The stationary phase precipitated within the packing remains there, while the extra-particle emulsion is eluted. Quite heavy loadings of stationary phase may be deposited in this way.

6) Column packings. Two main types of packing material are available. Those which are homogeneous and usually porous throughout, and those which consist of a solid core, usually glass, surrounded by a thin layer of partitioning or support material. The latter are usually called pellicular or surface layered packings.

For theoretical reasons, outlined in Chapter 4, a uniform homogeneous material is expected to give poorer performance, in terms of speed of resolution with a given pressure drop, than a pellicular material with the same composition. However, pellicular materials have a lower sample capacity (typically five times lower) and in practice do not always achieve the predicted performance.

Homogeneous packings based upon silica and alumina have long been used in adsorption and partition chromatography, and more recently in exclusion chromatography. Silica gels and porous glasses, which are roughly equivalent, although made by different processes, are available in chip and spherical forms with particle sizes from 5 to 50 μm, and with pore diameters from 5 to 500 nm. Aluminas are commonly available in chip form but recently a spherical alumina of excellent chromatographic properties has become available in particle sizes from 5 to 20 μm. All these materials can be used directly for adsorption chromatography after suitable conditioning with a solvent partially (say 50%) saturated with water, or they can be coated with a stationary liquid phase for partition chromatography. Wide pore silica gels and porous glasses are particularly suitable for liquid-liquid partition chromatography, and exclusion chromatography. 5—15 μm homogeneous adsorbents are becoming increasingly used and as noted above give very high performance in short wide columns (18, 19).

For ion-exchange chromatography, homogeneous resin beads or chips have long been used and the advantages to be gained from small particles is recognised in the most recent instruments which use 5—10 μm grades, but even so, the slow equilibration in ion-exchange resins severely limits the speed of separation by conventional ion-exchange chromatography.

Pellicular materials are now available for adsorption, partition and ion-exchange chromatography. The first materials were developed by Halasz and Horvath (20) and by Kirkland (21) for gas chromatography. Horvath, Preiss and Lipsky (7) in 1967 were the first to apply the pellicular idea to liquid chromatography but Kirkland (5) made the major break through when he used his material with dramatic effect in liquid-liquid partition chromatography. He also developed pellicular ion-exchangers (15) which were a substantial improvement on conventional ion-exchangers, and upon the first pellicular materials of Horvath, Preiss and Lipsky (7). Kirkland's materials with particle sizes around 30 μm gave very high speed analyses and have been shown (22) to be superior to all other materials for HSLC in terms of reduced plate height (see Chapter 4). These materials are sold by du Pont under the trade name Zipax. Other pellicular materials on the market generally show little advantage over homogeneous porous materials of the same particle size. Since 5—10 μm homogeneous particles now give as fast if not faster analyses than the best pellicular materials many of the initial arguments for using pellicular materials no longer apply.

A recent advance is the development of packings with the stationary phase chemically bonded to the support (11, 23). These materials are generally produced by reacting silanes, particularly chlorosilanes, with a silica surface to convert $\rightarrow$Si—OH into $\rightarrow$Si—O—SiR$_3$ where R can be any organic group ranging from a non polar group such as octadecyl to a highly polar group such as phenyl sulphonic acid. Several surface modified materials, both homogeneous and pellicular are now available and further development may confidently be expected. Such materials are thermally stable and are ideal for use with gradient elution.

7) Gradient systems. As noted in section 2 on pressurisation systems the type of pump used to generate the high pressure determines which kind of gradient system can be used. With a low-volume reciprocating pump it is possible, if the connections and ancillaries are properly designed to use a system which generates the solvent program on the low pressure side of the pump. The solvent programming system may take two solvents in a ratio which is controlled and programmed, or it may select a series of solvents differing slightly in composition or polarity from a series of reservoirs as shown in Figure 5A.

Alternatively if pressurisation devices are used which pressurise large volumes of solvent (pressure intensifiers, syringe pumps, direct gas pressurisation systems) any solvent program must be generated under high pressure conditions.

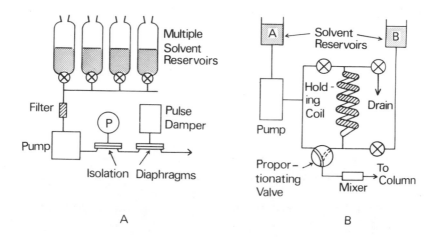

FIGURE 5. A) Low pressure gradient system. B) High pressure gradient system (after du Pont).

One such system (du Pont) (24) holds a second solvent in a coil of about 70 ml volume, as shown in Figure 5B. During operation solvent B in the holding coil is gradually displaced by solvent A with which the pump is filled. A and B are selected according to the predetermined programme by means of a proportionating valve with which cycles about 10 times per minute but with a variable opening time to channel A or channel B. A disadvantage of high pressure gradient systems is that only two solvents can be used for any run and they must of course be miscible. With a multi-pot low pressure system such as that shown in figure 5A a programme may be generated starting say with hexane and ending with water. As Scott has recently shown such a program can be achieved using 16 different solvents of successively increasing polarity (25).

CHAPTER 3

THEORY OF HIGH SPEED LIQUID CHROMATOGRAPHY
THERMODYNAMICS

The aim of high speed liquid chromatography is the rapid resolution of mixtures. Generally resolution is improved by reducing the elution velocity or by lengthening the column. However both these measures increase the time of analysis, and if both resolution and speed are to be improved other measures must be taken.

As shown in Figure 6, resolution may be improved either by increasing the

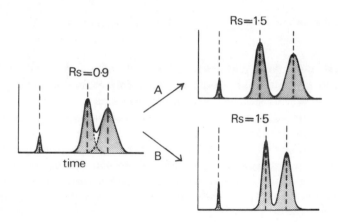

FIGURE 6. Improvement in resolution A, by increasing peak separation, and B, by reducing peak width. The first peak in each case represents an unretained solute.

separation of the eluted bands without changing their widths, or by reducing the peak widths without increasing their separation, or of course, by a combination of the two. Resolution, Rs, may be formally defined by equation (1) and the quantities in the equation are shown in Figure 7.

$$Rs = \frac{t_2 - t_1}{\frac{1}{2}(w_1 + w_2)} \tag{1}$$

t_1 and t_2 are the elution times of two adjacent peaks whose base widths are w_1 and w_2. The base width of a peak is the segment of baseline which is excised

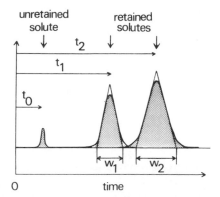

FIGURE 7. Definition of quantities in the equation for resolution (equation 1).

by extrapolation of the tangents to the points of inflection to the peaks. For a Gaussian peak whose concentration profile is defined by equation (2)

$$c = c_o \exp[-\Delta z^2 / 2\sigma^2] \qquad (2)$$

where Δz is the distance from the peak maximum, the base width equals four standard deviations, that is $w = 4\sigma$.

In general for well behaved chromatography, the condition for which is discussed below, the elution time of a solute relative to that of an unretained compound is determined entirely by thermodynamic considerations, while the peak width relative to the retention time is determined by kinetic considerations. Thus change of resolution be method A of Figure 6 involves changing the thermodynamic aspects of the chromatography, while changing resolution be method B involves improving the kinetics. These two aspects of chromatography are essentially independent and we therefore consider them separately, thermodynamics in this chapter, and kinetics in chapter 4.

Thermodynamic aspects

The relation of retention to equilibrium thermodynamic properties can be elucidated by considering a simple idealised experiment. A band of solute is eluted part way down a column and the solute flow arrested so that complete equilibration can take place between the solute in the mobile and stationary phases. The resulting situation is illustrated in Figure 8. The concentration profile of the solute band in the mobile phase is mirrored, apart from scale, by the profile in stationary phase, which for simplicity we take to be a liquid. Since the system is at equilibrium the concentration ratio is the equilibrium partition ratio for the solute between the two phases. This ratio is denoted by k and we assume that k is independent of the absolute values of the concentrations. The consequence of variation of k with concentration will be considered later. For any position in

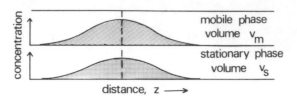

FIGURE 8. Concentration profiles in the mobile and stationary phase with arrested elution.

the concentration band we thus have:

$$c_s/c_m = k = \text{equilibrium partition ratio} \qquad (3)$$

where c_s and c_m are the concentrations (moles per unit volume) of solute in the stationary and mobile phases. The ratio, k', of the quantities, q_s and q_m, of solute in the two phases is called the column capacity ratio. For a column of uniform composition, k' is given by equation (4).

$$k' = \frac{q_s}{q_m} = \frac{c_s v_s}{c_m v_m} = k\,\frac{v_s}{v_m} \qquad (4)$$

v_s and v_m being the volumes of stationary and mobile phases in the column. Since all equilibrium is dynamically maintained, it follows that the quantity ratio, k', is equal within statistical limits, to the ratio of the times t_s and t_m, spent by typical molecules in the stationary and mobile phases. That is

$$k' = \frac{q_s}{q_m} = \frac{\overline{t_s}}{\overline{t_m}} \qquad (5)$$

and therefore that:

$$\text{Mean fraction of time spent by molecules in mobile phase} = \frac{\overline{t_m}}{\overline{t_m} + \overline{t_s}} = \frac{1}{1 + k'} \qquad (6)$$

Suppose now that the mobile phase is set moving at a linear velocity u. This movement cannot alter the average proportion of time spent by a typical molecule in the mobile phase. However when the molecule finds itself in the mobile phase it will move along the column at a speed u, whereas when it is in the stationary phase it is of course motionless. The mean downstream velocity of a typical molecule is thus given by:

$$u_{band} = u \times \begin{array}{c} \text{fraction of time spent} \\ \text{by typical molecule in} \\ \text{the mobile phase} \end{array} = \frac{u}{1 + k'} \qquad (7)$$

The relative migration rate, generally called the R_f-value, and which we denote for simplicity by R, is thus given by:

$$R = u_{band}/u = 1/(1 + k') \qquad (8)$$

whence

$$k' = (1 - R)/R \qquad (9)$$

equation (7) may be recast in terms of elution times since these are inversely proportional to band velocities:

$$t/t_o = u/u_{band} = 1 + k' \qquad (10)$$

t and t_o being the elution times of the retained and unretained bands. From equation (10) we immediately obtain:

$$k' = (t - t_o)/t_o \qquad (11)$$

k' is thus readily obtained from any elution chromatogram (see for example Figure 7) being equal to the net retention time of the band, $t - t_o$, divided by the time for elution of the unretained band. This explains why chromatographers familiar with elution techniques tend to use the capacity ratio more than the retention ratio R, while those familiar with in-column techniques (paper and thin layer chromatography for example) normally use R which is directly obtained from the concentration-distance chromatogram. k', as we have seen, is however more closely related to thermodynamic parameters, and is the preferred parameter for any theoretical treatment.

It is important to note that the different parts of a chromatographic band will move down the column at the same rate only if k and k' are independent of the absolute level of concentration. If for instance k' decreases with concentration as is often found in adsorption chromatography because of the non-uniformity of adsorption sites, the low concentration wings of the band possessing high values of k' will tend to move more slowly than the high concentration central sections. As the band moves down the column we therefore observe a 'breaking wave effect', the high concentration centre moving faster then the low concentration leading and trailing edges. The band thus becomes skew. The degree of skewing depends upon the range of k' over the concentration range within the band. If this is large the peak will broaden as it moves down the column more or less in proportion to the distance moved by the band. Thus increase in column length will not then improve the separation as much as would be expected if the band were symmetrical. Well behaved chromatography is therefore defined as chromatography under conditions when k' is independent of concentration.

Skewed peaks very often indicate that the stationary or mobile phases has been overloaded with sample, and the normal remedy is to reduce the sample size. In adsorption chromatography this may have little beneficial effect if the adsorption sites are very heterogeneous in activity. A simple way to effect improvement is to deactivate the strong adsorption sites by adsorption of a strongly held material such as water. Preconditioning of adsorbents is in fact generally necessary to ensure good peak shape.

In liquid-liquid chromatography the effects of overloading are likely to become noticeable when the concentration of solute in the stationary phase reaches about 3%. A rough rule of thumb for ensuring that overload does not occur is to restrict the total amount of solute to about 1/1000th of the total amount of stationary phase in the column. Since this will be around 100 mg for a 2mm bore 1m long column, overload may be anticipated when the sample size reaches about 100 μg. To prevent increased peak spreading beyond that expected from a infinitesimal sample size, the volume of solution injected should not exceed about 1/30th of the mobile phase volume in the column. For the typical column this means that sample volumes should not exceed about 70μl. With adsorbents there is no general rule because the point at which overload becomes a problem depends upon the shape of the adsorption isotherm. Experience shows however that for high surface area materials, say those with areas of about 100 m^2 g^{-1} overloading becomes significant under much the same conditions as those described for LLC (26).

The effect of temperature on retention is determined by its effect on k or k', and this may be related by standard thermodynamics to the heat of transfer $\Delta H_{s \to m}$ of the solute from the stationary to the mobile phase. The relationship is the Van't Hoff equation

$$d \ln k'/dT = -\Delta H_{s \to m}/RT^2 \qquad (12)$$

Generally in liquid chromatography heats of transfer are smaller than in gas chromatography and can in principle even be negative. However, for retention to occur to a significant extent the solute molecules must partition themselves strongly in favour of the stationary phase, since the ratio of stationary to mobile phase is normally much smaller than unity. Since entropies of transfer are likely to be small, $\Delta H_{s \to m}$ will normally be positive and values around 10 kJ mol^{-1} are common. These are about a quarter of the values found in GC. Because of the relatively small values of $\Delta H_{s \to m}$ temperature is not a particularly important parameter from the thermodynamic point of view in LC. The working temperature for LC operation is usually determined more by kinetic than thermodynamic considerations. Since diffusion coefficients and fluidities of liquids increase markedly with temperature it is advantageous to work at the highest possible temperature to optimise the kinetics. Once the best temperature from this point of view has been selected the thermodynamic aspects of the separation can be optimised by adjustment of the compositions of the mobile and stationary phases to give the optimum selectivity ratio a, where

$$a = k_2/k_1 = k_2'/k_1' \qquad (13)$$

Whereas the selection of the stationary phase in gas chromatography can now be made on a reasonably scientific basis, rules for the selection of optimum mobile and stationary phase combinations for LC are still rudimentary. In adsorption chromatography, as a result of the work of Snyder (27), the situation is reasonably well understood in a semi quantitative way. Broadly speaking for an adsorbent like silica, which may occur in a wide variety of forms from high surface area silica gels to low area porous glasses, the degree of adsorption, or k' value of a solute with a given mobile phase and adsorbent activity, is proportional to the surface area. Retention is chiefly determined by the ease with which solvent molecules on the surface can be displaced by solute molecules. The processes may be represented formally by the equilibrium equation (14)

$$X + nS_{ads} = X_{ads} + nS \qquad\qquad (14)$$

where X and S are solute and solvent molecules respectively, and n is the number of solvent molecules which one solute molecule must displace on adsorption. The more easily solvent is replaced by solute the more strongly is the solute retained. If the solvent is more strongly adsorbed than the solute and cannot easily be replaced at the surface, solute molecules may be excluded from exchange with adsorbed solvent and so can be eluted before typical solvent molecules. This type of exclusion may be important in molecular exclusion chromatography in addition to exclusion based upon molecular size.

Synder (27) has arranged the common solvents used in adsorption chromatography in an eluotropic series in which each solvent has a stronger eluting power than the one above it in the list. According to Synder's scheme each solvent is given an ϵ^0 value which indicates its eluting power. Some typical values are listed in Table 1. The ϵ^0 value of a solvent is formally related to its free energy of adsorption, but in practice ϵ^0's are determined empirically from the values of log k' for a number of standard solutes.

TABLE 1

Selected Eluotropic Series of UV Transparent Solvents for Alumina

from Synder (27)

Solvent	ϵ^0	η^{**}	Solvent	ϵ^0	η
pentane	0.00	0.23	tetrahydrofuran	0.45	—
cyclohexane	0.04	1.00	ethylene dichloride	0.49	0.79
l-pentene	0.08	—	dioxan	0.56	1.54
carbon tetrachloride*	0.18	0.97	amyl alcohol	0.61	4.1
di n-propyl ether	0.28	0.37	dimethylsulphoxide	0.62	2.24
propyl chlorides	0.29	0.33	acetonitrile	0.65	0.37
ethyl bromide	0.37	—	2-butoxyethanol	0.74	—
ethyl ether	0.38	0.23	propanols	0.82	2.3
chloroform	0.40	0.57	ethanol	0.88	1.2
methylene chloride	0.42	0.44	methanol	0.95	0.60
			acetic acid	high.	1.26

*   The UV cut off for carbon tetrachloride occurs at 265 nm.

** Viscosities, η, in centipoise for 20°C.

To establish the conditions for elution of the components of an unknown mixture, several solvents are used with increasing ϵ^o values until the components of the mixture are eluted with reasonable k' values say between 0.3 and 10. With experience the range of ϵ^o values which should be used can be selected quite easily. In the final selection of the optimum solvent, it is necessary to examine several solvents with roughly equal ϵ^o's. Synder has recently developed a procedure which can be used to remove some of the empiricism from this final selection (28).

When one comes to liquid-liquid chromatography somewhat similar considerations apply with the difference that there is now a very wide range of possible stationary phases. A highly polar stationary phase, for example a cyano-ether, would be used with a non-polar mobile phase such as hexane to give the longest retention time for a group of polar solutes. The use of a more polar mobile phase such as dibutyl ether would reduce the elution time. In chromatographing hydrophobic compounds one might use a non-polar stationary phase such as squalane and a polar mobile phase. The retention volume would be reduced by reducing the polarity of the mobile phase. As we have already noted the range of solvents and stationary phases which can be used for LLC is seriously curtailed by the requirement that the two phases must be mutually rather insoluble. The problem is high-lighted by noting that the total loading of stationary phase in a LLC column may be only about 100mg, while the amount of solvent which one might reasonably expect to pump through the column during its useful life might amount to several litres. To maintain a partition column in a stable condition it is essential to ensure that the mobile phase is accurately saturated with stationary phase before it enters the analytical column. This can be achieved satisfactorily only by using a precolumn which is maintained at exactly the same temperature as the analytical column. The precolumn is packed with an inexpensive support heavily loaded with stationary phase.

The prediction of retention times or k' values for LLC systems is still at an early stage. But the signs are hopeful that a framework will be developed, and a number of attempts have been made to rationalise the subject (29,30).

Supports bearing chemically bonded stationary phases can be treated much as adsorbents, although their thermodynamic behaviour is closer to that of liquids. They possess the great advantage that the stationary phase is not removed by passage of solvents of similar chemical composition. It is thus possible to use a solvent program starting with water or methanol and gradually changing to hexane, with a bonded hydrocarbon stationary phase such as ODS Permaphase. In this way almost any mixture can be eluted whatever the relative polarities of its constituents. The ability to use a solvent like hexane with a hydrocarbon stationary phase means that the support can always be cleaned of any strongly retained material. This can be a problem in normal adsorption or liquid–liquid chromatography.

In ion-exchange chromatography the pH and ionic strength of the solvent are the key parameters (31), but the nature of the ions present in the eluent and its content of miscible organics may have a significant effect on separation. These latter effects are difficult to interpret. In principle, however the effects of pH and ionic strength can be treated semi-theoretically.

Ion-exchange resins consist of cross linked polymers bearing frequent ionic groups. The commonest are the sulphonic acid and quaternary ammonium groups. These groups give respectively strong cation and strong anion exchange resins. Under normal operating conditions with an electrolyte as mobile phase, strong ion-exchangers possess a fully ionised skeleton with counter ions closely associated with each centre. We may thus represent these resins as follows:

Strong cation exchanger $\sim\sim\sim SO_3^-$, Na^+

Strong anion exchanger $\sim\sim\sim NR_3^+$, NO_3^-

In weak ion-exchange resins the functional groups are carboxyl, or amino groups whose degree of ionisation will depend upon the pH of the mobile phase. Under the conditions normally useful for chromatography the counter ions exchange with ionised solute molecules. In a typical sulphonated polystyrene there will be a sulphonic acid group for nearly every styrene group. Thus the ion-exchange groups are present throughout the polymer and it is important that they are readily accessible to the solute ions if efficient chromatography is to be achieved. Poor accessibility is one of the main limitations of present ion-exchange materials and can lead to excessive band spreading because of the poor mass transfer rates which result. This problem is partially overcome by the use of micro-porous resins and pellicular materials.

The role of pH of the eluent on retention in ion-exchange chromatography can be understood by consideration of simple ionic equilibrium. With an amine in aqueous solution for example, which might be chromatographed on a cation exchanger, the following equilibrium exists.

$$RNH_2 + H_2O \rightleftharpoons RNH_3^+ + OH^- \tag{15}$$

The equilibrium constant, K_b, the base dissociation constant, is

$$K_b = [RNH_3^+] [OH^-]/[RNH_2] \tag{16}$$

and by elementary equilibrium theory it follows that

$$Log [RNH_3^+]/[RNH_2] = 14 - pH - pK_b \tag{17}$$

where $pK_b = -Log_{10}K_b$, and the number 14 is the log of the ionic product of water. Since it is the ionic form of the amine which exchanges with the counter ion of the resin, increase of pH will decrease the retention since the ratio of ionised to neutral amine decreases.

Ionic strength affects not the dissociation of the base, but the degree of retention of the ionised form of the base. This may be understood in terms of the exchange equilibrium:

$$\sim SO_3^-, Na^+ + RNH_3^+ = \sim SO_3^-, RNH_3^+ + Na^+ \qquad (18)$$

The situation is analogous to that with an adsorbent as can be seen by comparing equations (14) and (18), where the solute molecule or ion has to displace a solvent molecule or ion from the adsorption or ion-exchanging site. As in adsorption chromatography where the degree of retention is altered by changing the solvent, so in ion-exchange chromatography the degree of retention can be altered by changing the counter ion. Qualitatively these predictions agree with experiment for increase of ionic strength decreases the degree of retention. Thus simultaneous changes in pH and ionic strength can effect the kind of changes in resolution illustrated by A of Figure 6.

The presence of soluble organic substances in the eluent may have an important effect on the dynamics of chromatography by causing swelling of the resin and thus rendering the ionic sites more accessible to solute ions. Organic solvents will furthermore be preferentially absorbed by the resin, and this will alter the nature of the resin phase by making it more organophilic. Unionised forms of the solutes will now be more strongly retained while the ionised forms will be less strongly retained. Whether an amine (in the case of a cation exchange resin) is more or less strongly retained will be difficult to predict for the two factors tend to cancel. In practice alcohols added to the mobile phase are found to decrease the retention of amines showing that the second effect predominates.

With anion exchange resins similar considerations apply. Increase of pH will increase rather than decrease retention but changes in the other parameters considered will have similar effects.

CHAPTER 4

THEORY OF HIGH SPEED LIQUID CHROMATOGRAPHY - KINETICS

If we enquire more closely into what exactly occurs when the the flow is restarted after arresting elution and allowing the system to equilibrate (Figure 8) we note that the concentration band in the mobile phase must inevitably move slightly ahead of the associated band in the stationary phase thereby producing a slight displacement Δz. The situation is shown in Figure 9.

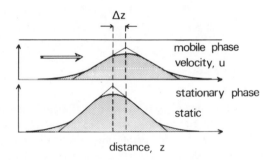

FIGURE 9. Concentration profiles during elution chromatography in the mobile and stationary phases.

The tendency for the two profiles to be pulled apart by the flow of mobile phase against stationary phase is countered by the equilibration process which continually tries to bring the concentrations in the mobile and stationary phase towards their equilibrium values. It is not difficult to show that the displacement Δz is simply related to the relaxation time for equilibration, τ, and the mean flow rate of the mobile phase, u, by equation (1)

$$\Delta z = u\tau \tag{1}$$

Thus the lower velocity and the faster equilibration, the smaller is Δz. Figure 9 also shows that the concentration or capacity ratio upstream of the peak maximum concentration is higher than the equilibrium ratio, and that the downstream ratio is lower than the equilibrium ratio, that is

$$(c_s/c_m)_{upstream} > k'_{equilib} > (c_s/c_m)_{downstream} \tag{2}$$

whence

$$k'_{upstream} > k'_{equilib} > k'_{downstream} \qquad (3)$$

By equation (3.6) we then deduce that:

$$u_{band,\,upstream} < u_{band,\,centre} < u_{band,\,downstream} \quad (4)$$

The mean velocity of molecules upstream of the band centre is less than that of molecules in the centre, and this in its turn is less than the velocity of molecules further downstream. In other words, the band continually broadens as it moves down the column. It is unfortunately not possible in general to measure Δz, and it must be inferred from the rate of spreading of the band as it moves along the column. When the band is Gaussian and is initially sharp it can be shown (9) that for small departures from equilibrium (i.e. Δz small with respect to the band width) the variance of the peak is directly related to the distance moved, that is

$$\sigma_z^2 \;=\; 2(1-R)\,\Delta z\, z \qquad (5)$$
$$=\; 2(1-R)\, u\tau\, z$$
$$\text{or} \quad \sigma_z^2/z = 2(1-R)\, u\tau \qquad (6)$$

This result is of great significance since it shows that the peak width $w = 4\sigma$ increases as the square root of the distance migrated. Since the separation of any two bands increases linearly with z, the resolution of peaks (see Equation 3.1) increases with the root of the distance travelled. This explains why resolution is increased by increase of column length. Where peak dispersion results from thermodynamic effects, such as non-linear isotherms, however, resolution will increase more slowly with migration distance. For efficient or well behaved chromatography it is vital to ensure that partition or adsorption isotherms are linear and peaks symmetrical.

Equation (5) also shows that band spreading is rooted in the balance of non-equilibrium, and that efficient chromatography is possible only when the equilibration process is rapid. Chromatography is a near equilibrium process and its improvement is almost exclusively the result of improvement in equilibration rates within and between the mobile and stationary phases.

Because of the clear connection between the increase of peak variance and the distance migrated by the band, the quantity σ_z^2/z or the differential form $\partial\sigma_z^2/\partial z$, which has the dimensions of a length within the column, is used as the general measure of column efficiency and is called the 'height equivalent to a theoretical plate' or more simply the plate height, H.

$$H \;=\; \partial\sigma_z^2/\partial z \qquad (7)$$

or for a uniform column of length L,

$$H = \sigma_L^2/L \qquad\qquad (8)$$

If the retention time of a peak is t_R and its variance in time units is σ_t^2, the plate height is given by

$$H = (\sigma_t/t_R)^2 L = (1/16)(w_t/t_R)^2 L \qquad\qquad (9)$$

The plate height is normally calculated from this formula in elution chromatography.

The concept and terminology of the plate height arose from the original paper of Martin and Synge in 1941 (3), in which they recognised for the first time that chromatography could be compared to a multi-stage counter-current extraction procedure analogous to distillation . They showed that the chromatographic process could be reproduced by a model consisting of a stack of plates each of such a thickness that the composition of the eluate emerging from any plate was the same as the concentration which would exist in the mobile phase over the whole plate if it were isolated and allowed to equilibrate. The plate theory of chromatography showed that an initially sharp band would develop a Gaussian concentration profile after a sufficiently long elution, that the variance of the profile obeyed equation (7), and that the rate of migration of the peak maximum obeyed equation (3.6).

Since 1941 the plate height has become the prime measure of column efficiency (or more correctly column inefficiency). It is readily visualised as a distance within the column, and it is simply related to the non-equilibrium displacement Δz by equation (6) which may now be written

$$H = 2(1-R)\Delta z = 2(1-R)u\tau \qquad\qquad (10)$$

When $R = 0.5$, $H = \Delta z$. The number of plates, which measures the efficiency of the column as a whole, is given by equation (11)

$$N = L/H = 16(t_R/w_t)^2 \qquad\qquad (11)$$

A 1600 plate column will give a peak whose base width is a tenth of the elution time. This represents an acceptable efficiency. Generally satisfactory columns will be equivalent to between 1000 and 5000 theoretical plates.

The crux of the non-equilibrium theory of chromatography is the calculation of the time constant, τ, for the relaxation of the non-equilibrium generated by the flow of mobile phase. The evaluation is by no means simple, except in special cases, and the problem has not been completely solved. However the work of Giddings (9) who has done most to develop the non-equilibrium theory has gone a long way to elucidating the important parameters affecting τ .

In general there are several sources of non-equilibrium in a real chromatographic column, and generally these different constituents of the total non-

equilibrium must be relaxed independently. When this is so, the total time constant for relaxation is the sum of the time constants for the individual relaxation
processes. Three independent relaxation processes are generally recognised, and
we can think of them as penetrating progressively deeper into the structure of
the column packing:

(1)   relaxation of the non-equilibrium generated by the difference in velocity within the flowing part of the mobile phase in the inter-particle space. This
non-equilibrium can be relaxed by the co-operative operation of the flow itself
and transverse|diffusion. The flow takes molecules from fast to slow moving
regions because of its tortuous nature while diffusion across the stream lines can
produce the same effect.

(2)   relaxation of the non-equilibrium generated by the flow of the inter-
particle mobile phase over the stagnant part of the mobile phase held in the interstices of the packing material. This is relaxed only by diffusion in and out of the
porous particles, there being no flow within the particles.

(3)   relaxation of the non-equilibrium generated by the flow of mobile
phase as a whole over the stationary phase. This is relaxed by diffusion in and
out of the stationary phase if a liquid, or by desorption if an adsorbent.

When the relaxation process involves only diffusion, as in (2) and (3), the
time constant is proportional to the diffusion time which is given by the Einstein
equation:

$$\tau_{diff} = d^2/2D \tag{12}$$

where d is the root mean square distance to be diffused by a typical molecule,
and D is the diffusion coefficient. Since it is important to reduce τ to a minimum
it is important to ensure that the distances over which diffusion occurs are as
small as possible and diffusion coefficients are as large as possible. The diffusion
distances are respectively the thickness of the porous layer which contains the
stagnant mobile phase, and the thickness of the stationary phase film. For a
completely porous support the effective 'porous layer thickness' is about 1/3rd
of the particle diameter.

The time constant for the relaxation of the disequilibrium generated by the
complex flow pattern in the flowing mobile phase is more difficult to derive.
Indeed there is no theory which gives a quantitatively correct interpretation of
the experimental data. When two processes co-operate to relax a deviation from
equilibrium the rate constant for the overall process is the sum of the rate constants for the co-operating processes. Application of this idea in its simplest form
leads to the coupled equation of Giddings (9) for the contribution to H from the
flowing part of the mobile phase.

$$H = \frac{1}{\dfrac{1}{A} + \dfrac{1}{C\,u}} \tag{13}$$

This equation gives a poor representation of the actual dependence of H upon linear velocity for an unretained solute eluted from a column of imprevious particles (eg glass beads) (32). The dependence can however be reasonably well expressed (17) by a modified form of equation (13):

$$H = \frac{1}{\frac{1}{A'} + \frac{1}{C'u^{n'}}} \quad ; \quad n' \approx 0.3 \tag{14}$$

For a ten fold range of u equation (14) is adequately approximated (22) by the simpler equation (15)

$$H = Du^n \tag{15}$$

The contributions of relaxation processes (2) and (3) to the plate height have been evaluated by Giddings (9) and the total plate height contribution from non-equilibrium can be written

$$H_{ne} = D u^n + q_s \left\{ \frac{k'}{(1+k')^2} \right\} \frac{d_f^2}{D_s} u + q_m \left\{ 1 + \frac{ak'}{(1+k')} \right\}^2 \frac{d_l^2}{D_m} u \tag{16}$$

flow	stationary	stagnant mobile
contri-	phase	phase contribution
bution	contribution	

where q_s, q_m and a are configurational factors reflecting the geometry of the stationary and stagnant mobile phase; d_f and d_l are the thickness of the stationary phase and porous layer containing the stagnant mobile phase; and D_s and D_m are the diffusion coefficients of solute in the stationary and mobile phases.

In addition to peak dispersion from slow relaxation of the non-equilibrium generated by the flow, dispersion also results from axial molecular diffusion. This may occur either in the mobile or stationary phase, and contributions to H is given by equation (17)

$$H_{diff} = (2\gamma D_m + 2\gamma' D_s k')/u \tag{17}$$

Here γ and γ' are obstructive factors for diffusion in the mobile and stationary phases respectively. In GC the second part of the diffusion term can be neglected since D_s is about 10^4 times smaller than D_m, and k' is normally not larger than 10; in LC, however, the second term cannot be ignored since both diffusion coefficients are likely to be comparable, Fortunately for most operating conditions in high speed liquid chromatography the flow speed is such that the axial diffusion may be ignored. This may not however be always the case, for as we move towards the use of smaller and smaller particles this term will become more important.

The total plate height is then the sum of the non-equilibrium and diffusional contributions. When comparing different materials in liquid chromatography and when comparing analyses under different conditions of fluid velocity and with different eluents it is desirable to be able to compare experimental data on the basis of some common yardstick. As shown by Giddings (9, 33) the most suitable parameters for such comparisons are dimensionless parameters called the reduced plate height and the reduced fluid velocity. These are defined by equations (18) and (19).

$$\text{Reduced plate height,} \quad h = H/d_p \tag{18}$$

$$\text{Reduced fluid velocity,} \quad \nu = ud_p/D_m \tag{19}$$

The reduced plate height can be thought of as a plate height scaled to the particle diameter, and the reduced velocity as the rate of flow compared to the rate of diffusion over a particle. Reduced parameters provide a particularly simple method for optimising performance in chromatographic systems. When the plate height-velocity equation is expressed in reduced terms considerable simplification results and the new equation contains only ratios of distances and of diffusion coefficients. Setting n of equation (15) equal to 0.33, the total reduced plate height may be expressed in the form

$$h = (\gamma + k'\gamma' \frac{D_s}{D_m}) \frac{2}{\nu} + A \nu^{0.33}$$

$$+ q_m \left\{ 1 + \frac{ak'}{1+k'} \right\}^2 \left| \frac{d_l}{d_p} \right|^2 \nu$$

$$+ q_s \left\{ \frac{k'}{(1+k')^2} \right\} \left| \frac{d_f}{d_p} \right|^2 \frac{D_m}{D_s} \nu \tag{20}$$

For practical purposes this equation may be simplified (21) to

$$h = B/\nu + A\nu^{0.33} + C\nu \tag{21}$$

This equation is particularly useful since it is readily tested experimentally. Values of A and C are readily found and B can be calculated within a factor of better than two from estimates of the obstructive factors and diffusion coefficients and the known k'. Since A reflects the dispersive power of the interparticle space, while C reflects the dispersive power of the stationary and stagnant mobile phase we can readily test whether a column is well or badly packed (low or high values of A) and whether it has good or bad mass transfer characteristics (low or high C). Generally a well packed column will have a value of A around unity and one with good mass transfer characteristics will have a C-value below 5×10^{-2}. γ is about 0.6 for impervious packings and closer to unity for porous packings γ' is likely to be near unity.

The way in which the various terms of equation (21) contribute to the total plate height is illustrated in Figure 10.

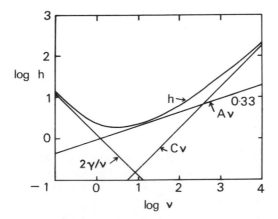

FIGURE 10. Contributions to h from the various terms of equation (21). Assumed values B=2, A=1, C=2×10^{-2}.

The curve shows a flat minimum at a reduced velocity around 3 and a reduced plate height just above unity. In gas and thin layer chromatography it is usual to operate at reduced velocities near the optimum as shown in Table 2. In current HSLC much higher reduced velocities are used, generally between 100 and 1000.

TABLE 2

Comparison of operating conditions in GC and LC

Mode	u	D_m	d_p	ν	h*
Gas Chromatography	10	0.1	2×10^{-2}	2	1-3
Thin Layer Chromatography	2×10^{-2}	10^{-5}	10^{-3}	2	5-10
High Speed Liquid Chromatography	2	10^{-5}	3×10^{-3}	600	10-30

* Reasonable values in practical operation. Units throughout are cgs.

Equations (20) and (21) can be used as a yardstick for comparing different chromatographic packings under a very wide range of operating conditions. Knox and Saleem (34), for instance showed that identical (h, ν) curves were obtained when a given column was used for gas and liquid chromatography although the diffusion coefficients D_m differed by four orders of magnitude. A more restricted type of comparison is one in which data are available on a given packing material using different solvents. These can most readily be compared on a reduced plate height basis since this allows for differences in the diffusion coefficients. When it is wished to compare columns with different particle size one normally finds that the value of H obtained with the smaller particles is itself smaller, but one is often more interested to know if the improvement found with the smaller particles is as great as might be expected theoretically. This is readily determined from reduced plots since it is predicted by the theory, embodied in equation (20), that identical (h, ν) plots will be obtained for equally well packed columns of a given material whatever the particle size. Usually it is found in practice that the (h, ν) plots for smaller particles lie above those for larger particles because of packing problems, and that the A-values are higher.

Very often experimental data exists over a rather restricted range of ν and the plot of log h against log ν is straight within experimental error, that is

$$h = d\nu^n \tag{22}$$

where n will lie between 0.33 and 1.0. If $\nu > 10$ axial diffusion may be neglected and only the last two terms in equation (21) are significant. When this is so the determination of n is sufficient to determine the proportion of h contributed by the terms $A\nu^{0.33}$ and $C\nu$. These proportions as a function of n are shown in Table 3.

TABLE 3

Contributions to h from terms in equation (19)

Value of n of equation (22)	Proportion of h contributed by	
	$A\nu^{0.33}$	$C\nu$
0.33	100	0
0.4	90	10
0.5	75	25
0.6	60	40
0.7	45	55
0.8	30	70
0.9	15	85
1.0	0	100

For example, suppose that a plot of log h against log ν gave n = 0.6, and that the median values on the log plot were h = 25 at ν = 300. From the Table we see that the A and C terms contribute 60% and 40% of h respectively: that is, $A\nu^{0.33}$ = 0.60 h = 15 and $C\nu$ = 10. Since ν = 300 we immediately obtain A = 2.25 and C = 3.3×10^{-2}. Such values of A and C indicate a moderately well packed column with reasonable mass transfer characteristics. In practice a well packed column of 20μm silica gel might give such values.

When the range of ν covered by the observations is large or the data very accurate, A and C are best found by curve fitting. Allowance is thereby made for the slight upward curvature of most log h versus log ν plots. Some typical plots are given in Figure 11, and the appropriate values of A and C are given in Table 4.

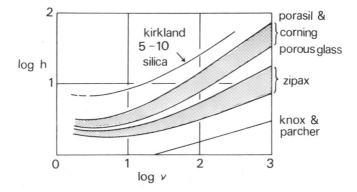

FIGURE 11. Reduced plate height, reduced velocity plots for typical supports used in high speed liquid chromatography.

TABLE 4

Values of A and C for Typical Supports used in
High Speed Liquid Chromatography

Material	k'	A	C	ref.
Zipax 29μm	0	0.75	0	22
	1.7	0.85	0.7×10^{-2}	22
	7.8	1.0	1.1×10^{-2}	22
Porasil 60 and 250	0	1.5	3.5×10^{-2}	22
50μm	0.4-0.7	1.7	5.0×10^{-2}	22
Corning Porous Glass	0	1.4	1.7×10^{-2}	22
50μm	0.4-4	1.7	4.5×10^{-2}	22
Micro-porous silica 6-10μm	>1	3.7	7×10^{-2}	18
480μm glass spheres	0	0.4	0	17

Plots such as those shown in Figure 11 or the A and C values of Table 4 enable further information to be obtained about the chromatographic charact-eristics of supports, and in certain cases suggestions can be made as to how to improve their chromatographic properties. This has been discussed in some detail by Done, Kennedy and Knox (22). The important points may be illustrated by reference to Figure 11, where the lower limits of the bands refer to unretained solutes while the upper limits refer to strongly retained solutes (k' large); single lines refer to single solutes.

The lowest line was obtained by Knox and Parcher (17) for a column con-taining 480 μm diameter glass beads. This is the lowest plate height curve ever obtained and corresponds to A = 0.4. The next lowest curve was obtained with fractions of Zipax ranging in diameter from 29 to 115μm. All fractions gave identical (h, ν) curves indicating that they were all equally well packed. The value of A was close to unity. With retained solutes the (h, ν) curves lie progressively higher as k' rises. This is expected from equation (20) when the mobile phase mass transfer term dominates and the stationary phase mass transfer term is small, for the latter passes through a minimum when k' = 1. Zipax is a pellicular material so d_l/d_p is small and therefore C is expected to be small. Its value is around 1×10^{-2}.

Porasil, unlike Zipax, is a completely porous material and even for an unretained solute C is not zero. A significant contribution to h arises from slow mass transfer in the stagnant mobile phase. Thus the plate height curve is both higher and steeper than that obtained for Zipax. A is only slightly larger than that for Zipax but C is much higher. C is only slightly dependent upon k'.

The curve for the 6-10μm silica is taken from the recent data of Kirkland (18). Similar data has been obtained by Majors (19) with 6-44μ silica gels. The curve is higher than that for the 8 times larger Porasil but only slightly so. The curve is flatter which indicates a rather high value of A at around 4 while C is only a little higher than for Porasil, 7×10^{-2} compared to 5×10^{-2}. The problem with the fine particle silica is therefore packing rather than mass transfer, and it may be possible to effect considerable improvements in performance by developing better packing methods.

For only one material so far studied in detail is there any evidence of limitation by slow mass transfer in the stationary phase, this being revealed by a maximum in h when k' is around unity. With Corasil, which consists of glass beads layered with silica gel, this behaviour is noted for certain classes of compound. The value of A is low, around 1.2 to 1.7, but C is very dependent on k' being zero for k' = 0 and around 5×10^{-2} for k' between 1 and 3; for lower and higher values of k' the C-value is lower. The precise reason for this behaviour is not clear, but it may be that the mass transfer characteristics of the silica gel layer have not been optimised.

At the present time only a few supports have been sufficiently well examined for their packing and mass transfer characteristics to be properly established, and undoubtedly much work remains to be done. The approach just given differs from that advocated elsewhere (18) but appears to give a useful framework within which to work for the improvement of column packings.

CHAPTER 5

CONCLUSIONS AND FUTURE DEVELOPMENTS

The illustrative chromatograms which follow in Part 2 show the great advances which have recently been made in the speed, sensitivity and resolution attainable in liquid chromatography as a result of the development of sophisticated LC equipment and of column packing materials. There is however still room for further advance and the directions in which this will occur are reasonably clear.

One of the first requirements for high speeds in LC was the generation of low flow rates of liquids under high pressures. Current equipment operates at pressures up to about 5000 psi (350 atm). There will be some advantage to be gained from proceeding to higher pressures but serious problems will eventually arise. Most connections used in current HSLC equipment are rated to 5000 psi and may become unsatisfactory if the pressure limit is pressed much beyond this. If exceptionally high column pressures are found to be necessary in order to enable efficiencies in the region of 100,000 plates to become realisable in a reasonable time, it may be desirable to design equipment to operate to 50,000 psi or even more. At this level however, liquids are appreciably compressible, and their thermodynamic properties differ from those at low pressure in ways which cannot be predicted. They also become more viscous so that some of the advantage expected from the high pressure is lost. Ultimately the energy generated by the frictional work as the fluid flows through the column will cause temperature gradients to develop across the column, and this will seriously impair resolution. It therefore seems unlikely that operating pressures will be much increased except for very special applications.

Considerable improvements in speed can still be obtained by employing smaller particles which enable faster equilibration to take place. The optimum particle size which will enable us to work at the minimum in the (h, v) curve with current pressure limits is between 2 and 3 μm (37), but the full benefit from the use of microparticles will only follow if they can be packed as well as particles ten times larger in diameter. If this could be achieved analyses which are currently carried out in minutes could be carried out in seconds. The recent work of Kirkland (18) and Majors (19) has shown that this limit is being approached quite rapidly for their 5 to 10 μm particles give substantially faster resolution than 30 μm particles. In going to microparticles, however, severe design problems will have to be faced since all dimensions must follow those of the column packing. Thus detector cells will have to be 0.1 mm in diameter instead of 1 mm, and other components will have to be reduced in size correspondingly.

Improvements in detectors will undoubtedly be made and are urgently required. At present the UV photometer and fluorometer are the only detectors

37

with adequate sensitivity for the analysis of trace components, but they are limited to UV absorbing compounds. We particularly need sensitive detectors which will respond to compounds having no UV absorption. It is generally agreed that some form of high sensitivity transport detection system will eventually be developed, but that this will be both difficult and expensive. Fluorometric detectors have obvious advantages over absorption detectors in that the signal from the solvent alone is zero rather than 100%, but they are still selective to UV absorbers. They do however offer great promise when coupled with microreactor systems which can be used to convert solutes containing a specific functional group into fluorescent derivatives. This seems an obvious field for exploitation.

Bulk property detectors on the other hand apparently have little ultimate potential except in restricted areas where high sensitivity is not required. Unfortunately the bulk properties of liquids are very little affected by small changes in composition, much less so than the bulk properties of gases where a large molecule replaces one, not several small eluent molecules. It is difficult to see how the limit of detection by such devices can be pushed much beyond 1 part in 10^7. What we require are detectors based upon selective principles, preferably rather broad ones, which can give detection limits in the region of 1 part in 10^{10} or 10^{11}.

Preparative HSLC will inevitably be developed within the next few years as was preparative gas chromatography, but the direction of development may well be different. The work of de Stefano and Beachell (16) shows clearly that with sophisticated column packing materials there is no decline in column performance when very wide columns are used if the samples are injected or removed in such a way that the material collected never contacts the walls. If the problems associated with injection and take-off can be solved there is no reason why highly efficient separations should not be carried out with columns which are as wide as they are long. Current columns can be used with loads as high as 100μg per mm^2 cross section without serious overloading being observed. A 10 cm diameter column should thus be able to handle about a gram of material.

The quantitative accuracy of HSLC has already been established by a number of workers to be as good as that of gas chromatography and the technique will undoubtedly be used more and more for quantitative work. One forsees that its use for plant monitoring and control will quickly become commonplace, and that it will be only a few years until nearly all chemical laboratories will use liquid chromatographs in much the same way as they now do gas chromatographs.

References

1. Tswett, M., *Trav. Soc. Nat. Warsovie,* **14,** (1903).
    English translation obtainable from W. Woelm, Eschwege, W. Germany.
2. Kuhn, R., Lederer, E., *Naturwiss,* **19,** 306 (1931).
    Kuhn, R., Lederer, E., Winterstein, A., *Ber. Deutsch. Chem. Gesell.,* **64,** 1349 (1931).
3. Martin, A.J.P., Synge, R.L.M., *Biochem, J.,* **35,** 1358 (1941).
4. James, A.T., Martin, A.J.P., *Biochem. J.,* **50,** 679 (1952).
5. Kirkland, J.J., *J. Chromatog. Sci.,* **7,** 7 (1969).
6. Huber, J.F.K., Hulsman, J.A.R.J., *Anal. Chim. Acta,* **38,** 305 (1967).
    Huber, J.F.K., *J. Chromatog. Sci.,* **7,** 85 (1969).
7. Horvath, C., Preiss, B., Lipsky, S.R., *Anal. Chem.,* **39,** 1422 (1967).
8. Hamilton, P.B., Bogue, D.C., Anderson, R.A., *Anal. Chem.,* **32,** 1782 (1960).
9. Giddings, J.C., *'Dynamics of Chromatography Part 1',* Marcel Dekker, New York 1965.
10. Kirkland, J.J., Ed. *'Modern Practice of Liquid Chromatography',* John Wiley, New York 1971.
11. Kirkland, J.J., de Stefano, J.J., *J. Chromatog. Sci.,* **8,** 309 (1970).
    Kirkland, J.J., *J. Chromatog. Sci.,* **9,** 206 (1971).
12. Halasz, I., Sebastian, I., *Angew. Chem. Int. Edn.,* **8,** 453 (1969).
13. Huber, J.F.K., *J. Chromatog. Sci.,* **7,** 172 (1969).
14. Scott, R.P.W., Lawrence, J.G., *J. Chromatog. Sci.,* **8,** 65 (1970).
15. Kirkland, J.J., *J. Chromatog. Sci.,* **7,** 361 (1969).
16. de Stefano, J.J., Beachell, H.C., *J. Chromatog. Sci.,* **8,** 434 (1970); **10,** 655 (1972).
17. Knox, J.H., Parcher, J.F., *Anal. Chem.,* **41,** 1599 (1969).
18. Kirkland, J.J., *J. Chromatog. Sci.,* **10,** 593 (1972).
    Kirkland, J.J., 163rd National Meeting, American Chemical Society, Chromatography Award, Symposium, Boston, April 1972.
19. Majors, R.E., *J. Chromatog. Sci.,* **11,** 88 (1973).
20. Halasz, I., Horvath, C., U.S. Patent 3,340,085; British Patent 1,016,635.
21. Kirkland, J.J., U.S. Patent 3,505,785.
22. Done, J.N., Kennedy, G.J., Knox, J.H., *'Gas Chromatography 1972'* Ed. Perry, Applied Science Publishers, London 1973, p. 145.
23. Locke, D.C., *J. Chromatog. Sci.,* **11,** 120 (1973).
24. Byrne, S.H., Schmit, J.A., Johnson, P.E., *J. Chromatog. Sci.,* **9,** 592 (1971).
25. Scott, R.P.W., *'Gas Chromatography 1972',* Ed. Stock, Institute of Petroleum, London 1973, p. 295.
26. Snyder, L.R., *J. Chromatog. Sci.,* **10,** 187 (1972).
27. Snyder, L.R., *'Principles of Adsorption Chromatography',* Marcel Dekker, New York, 1968.
28. Snyder, L.R., *'Gas Chromatography 1970',* Ed. Stock and Perry, Institute of Petroleum, London 1971, p. 81.
29. Huber, J.F.K., Meijers, C.A., Hulsman, J.A.R.J., *Anal. Chem.,* **44,** 111 (1972).
30. Keller, R.A., Karger, B.L., Snyder, L.R., *'Gas Chromatography 1970',* Ed. Stock and Perry, Institute of Petroleum, London 1971, p. 125.
31. Helferich, F., *'Ion Exchange',* McGraw Hill Book Co., New York 1962.
32. Knox, J.H., *Anal. Chem.,* **38,** 243 (1966).
33. Giddings, I.C., *Anal. Chem.,* **35,** 1338 (1963).
34. Knox, J.H., Saleem, M., *J. Chromatog. Sci.,* **7,** 745 (1969).
35. Kirkland, J.J., *J. Chromatog. Sci.,* **10,** 593 (1972).
36. Majors, R.E., *Anal. Chem.,* **44,** 1722 (1972).
37. Knox, J.H., Saleem, M., *J. Chromatog. Sci.,* **7,** 614 (1969).

PART 2

**HIGH SPEED LIQUID
CHROMATOGRAMS**

INTRODUCTION

The Chromatograms in Part 2 have been selected from the large number published over the last five years and obtained by the technique of high speed liquid chromatography described in Part 1. A particular separation selected may not be the best that has been achieved either in terms of chromatographic efficiency or selectivity for the technique is advancing so rapidly that improvements are continually being made. The most important recent advances are in the use of particles in the 5—20 μm range and a number of illustrative Chromatograms using microparticles are shown. It is, for example, probable that the separation of polyphenyls shown in Chromatogram 1.03 could be achieved several times faster using a finer grade of alumina than 20 μm, or possibly using a completely different partitioning system.

The Chromatograms have mainly been chosen to illustrate the great diversity of samples which have now been resolved, but also to illustrate the range of supports and stationary phases which have been used. A number of examples of the use of solvent programming or gradient elution have been included. Readers familiar with gas chromatography will recognise the close analogy between solvent programming in liquid chromatography, and temperature programming in gas chromatography.

The Chromatograms have been classified under eleven general headings:-
1. Hydrocarbons and Petroleum
2. Acids
3. Aromatic amines and other nitrogen compounds
4. Phenols and antioxidants
5. Sulphonic acids, dyes and dyestuff intermediates
6. Insecticides and herbicides
7. Nucleotides, nucleosides, and nucleic acid bases
8. Pharmaceutical products
9. Steroids
10. Natural products
11. Miscellaneous

The classification is self explanatory but admittedly somewhat arbitrary, and it has not always been easy to decide in exactly which section a particular Chromatogram should be placed, especially when it contains compounds belonging to more than one group. The classification of Chromatograms of acids was difficult but the large number of examples justified the two sections 2 and 5. Inevitably there were a number of separations containing acids which most naturally fell into section 4. Likewise it was not always easy to decide whether a separation fell more appropriately into section 8 or 10. The section titled Miscellaneous contains small groups of related separations, for example, aromatic

43

Chromatogram Number		Title

Chromatogram with time scale in minutes and
Peaks identified according to numbering on
Chromatogram.

Column material	Column length/m	Column bore/mm
Column packing	Particle size/μm	Temperature in $^{\circ}$C
Mobile phase and gradient		Stationary phase and percentage load
Inlet pressure/bar	Flow/cm s^{-1} or cm^3 min^{-1}	Sample size/μl
Detector	Operational details	Sensitivity

Additional comments

Reference and acknowledgment

esters and inorganic mixtures, but in the main contains Chromatograms of organic research samples and of various test mixtures designed to illustrate the potential of HSLC.

All compounds whose separations are illustrated in Chromatograms are listed in the name and formula index both under their trivial and where possible their full chemical names. Below each Chromatogram the appropriate literature reference is cited, and all authors mentioned in parts 1 and 2 are included in the author index. Appendix 1 gives details of the most widely used proprietary supports and packing materials with their manufacturers or suppliers.

Each Chromatogram is displayed in a standard format which gives a concise description of the operating conditions and other relevant data for the particular separation. The standard format contains a number of boxes, each box being used for detailing a specific operating parameter or piece of information. The designation of each box is shown in the format opposite.

The top line of the format gives the Chromatogram number followed by the title of the Chromatogram. A Chromatogram number such as 2.06 indicates the sixth Chromatogram of section 2 (acids). The title is intended as a concise description of the mixture being separated. The title of the section is given as the right hand page heading.

The Chromatogram of the mixture is placed in the largest box immediately below the title. Peak identification is by number, with the relevant compounds listed by the side or below the Chromatogram. All Chromatograms have the point of injection at the left, and so in many cases are the reverse of those in the original publication. All times are given in minutes from the moment of injection.

The first line below the box lists column parameters, the column material (glass, stainless steel, etc.), the column length (in metres) and the column diameter (in mm).

The second line gives details of packing; the material (often a proprietary name), the particle size (in microns, 1 micron = 10^{-6}m = 1 μm), and the column temperature (in degrees C).

The wider third line contains details of the mobile phase and solvent gradient (but see additional comments) and of the stationary phase (its constitution and loading given as a percentage by weight of the support). $\beta\beta'$-oxydipropionitrile, a very widely used stationary phase in liquid-liquid chromatography, has been denoted throughout by BOP.

The fourth line details the inlet pressure in bar: (1 bar = 10^5Nm^{-2} = 10^6 dyne cm^{-2} = 0.987 atm = 14.7 psi), followed by the flow rate. This is given as the linear velocity in cm s^{-1} as this gives the best indication of the speed of elution, but where the original paper gives insufficient data to allow it to be calculated, the volume flow rate is given in cm^3 min^{-1}. The last box gives the sample volume in microlitres (1 μl = 1 mm^3 = 10^{-3} cm^3).

The fifth line details the detector (uv, RI, wire transport, etc.), followed by any operating details such as the wavelength used for a photometric detector, and detector sensitivity. For u-v photometers the sensitivity is quoted as the absorbance for full scale deflection denoted by AUFS. Absorbance is defined by the equation:-

$$\text{Absorbance} = \log_{10}\left\{\frac{\text{Incident light intensity}}{\text{Transmitted light intensity}}\right\}$$

thus an absorbance of 0.01 absorbance units (AU) implies that 2.3% of the incident light has been absorbed.

The box labelled 'Additional Comments' is used to give fuller details of gradient systems, to explain packing methods and comment on separations of special significance. Descriptions of gradients can be complex and in the space available a complete description is not always possible. This is especially so when a home made gradient system is used and in such cases it may be possible only to give an outline of the system and to note whether the gradient (concentration versus time relation) is linear, concave or convex to the time axis.

The last lines in the format give the reference and acknowledgment.

1.01 POLYCYCLIC AROMATICS

1. 11H-Benzo[b] fluorene $C_{17}H_{12}$
2. 13H-Dibenzo[a,g]fluorene $C_{21}H_{14}$
3. Coronene $C_{24}H_{12}$

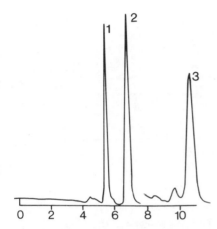

Stainless steel	1.00 m	2.1 mm
Zipax (R)	<37 μm	Ambient
Hexane		1.5% PEG 400
7 Bar	0.38 cm/sec	5 μl
Ultra-violet	254 nm	0.04 AUFS

Done, J.N., Knox, J.H., unpublished results

1.02 POLYCYCLIC AROMATICS

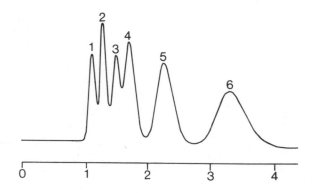

1. Decalin $C_{10}H_{18}$

2. Benzene C_6H_6

3. Naphthalene $C_{10}H_8$

4. Azulene $C_{10}H_8$

5. o-Quaterphenyl $C_{24}H_{18}$

6. m-Quaterphenyl $C_{24}H_{18}$

	1.00 m	2.3 mm
Alumina (Woelm)	38-53 μm	Ambient
Hexane		6% water
	2.9 cm^3/min	
Refractive index		

Bombaugh, K.J., Levangie, R.F., King, R.N., Abrahams, L., *Journal of Chromatographic Science,* **8**, 657 (1970). Reproduced by permission of Preston Technical Abstracts Co., Illinois.

1.03 POLYPHENYLS

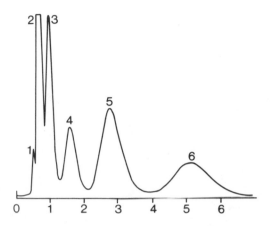

1. Benzene C_6H_6
2. Biphenyl $C_{12}H_{10}$
3. *m*-Terphenyl $C_{18}H_{14}$

4. *m*-Quaterphenyl $C_{24}H_{18}$
5. *m*-Quinquephenyl $C_{30}H_{22}$
6. *m*-Sexiphenyl $C_{36}H_{26}$

Steel	0.50 m	2.0 mm
Kieselgel (R)	30-40 μm	Ambient
Heptane		40% Fractonitril III
	0.83 cm/sec	
Ultra-violet	254 nm	

Randau, D., Schnell, J., *Journal of Chromatography,* **57**, 373 (1971). Reproduced by permission of Elsevier, Amsterdam.

1.04 SIMPLE AROMATIC COMPOUNDS

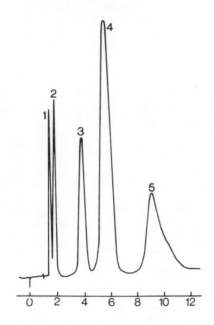

1. Benzene C_6H_6
2. Tetrahydronaphthalene $C_{10}H_{12}$
3. Styrene C_8H_8
4. Indene C_9H_8
5. Naphthalene $C_{10}H_8$

Copper	1.00 m	1.6 mm
Alumina (Woelm)	20-25 μm	22°C
Pentane		None

94 Bar	1.4 cm³/min	
Ultra-violet	254 nm	

Alumina deactivated by injection of 5 μl of thiophene

Martin, M., Loheac, J., Guiochon, G., *Chromatographia*, **5**, 33 (1972). Reproduced by permission of Friedr. Vieweg + Sohn GmbH, Braunschweig.

1.05 POLYCYCLIC AROMATICS

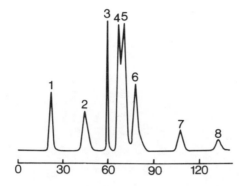

1. Indane C_9H_{10}
2. Naphthalene $C_{10}H_8$
3. Acenaphthene $C_{12}H_{10}$
4. Fluorene $C_{13}H_{10}$

5. Phenanthrene $C_{14}H_{10}$
6. Pyrene $C_{16}H_{10}$
7. Chrysene $C_{18}H_{12}$
8. Carbazole $C_{12}H_9N$

Glass	1.00 m	4.0 mm
Alumina (Woelm)		Ambient
Gradient of pentane to diethyl ether		None
	0.75 cm³/min	
Ultra-violet	260 nm	

Diethyl ether added to pentane at an increasing rate to give an approximately exponential gradient.

Popl, M., Mostecky, J., Havel, Z., *Journal of Chromatography,* **53**, 233 (1970).
Reproduced by permission of Elsevier, Amsterdam.

1.06 POLYCYCLIC AROMATICS

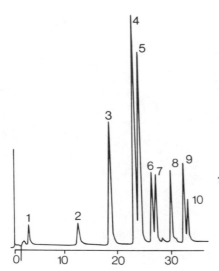

1. Benzene C_6H_6
2. Naphthalene $C_{10}H_8$
3. Biphenyl $C_{12}H_{10}$
4. Phenanthrene $C_{14}H_{10}$
5. Anthracene $C_{14}H_{10}$
6. Fluoranthene $C_{16}H_{10}$
7. Pyrene $C_{16}H_{10}$
8. Chrysene $C_{18}H_{12}$
9. Benzo[a] pyrene $C_{20}H_{12}$
10. Benzo[e] pyrene $C_{20}H_{12}$

Stainless steel	1.00 m	2.1 mm
Permaphase (R) ODS	<37μm	50°C
Gradient: 20% methanol in water to 100% methanol at 2% per minute		1% octadecylsilyl chemically bonded
70 Bar	1 cm³/min	
Ultra-violet	254 nm	

A good example of the use of gradient elution with a chemically bonded stationary phase to elute compounds of widely differing polarities

Schmit, J.A., Henry, R.A., Williams, R.C., Dieckman, J.F., *Journal of Chromatographic Science*, **9**, 645 (1971). Reproduced by permission of Preston Technical Abstracts Co., Illinois.

1.07 POLYCYCLIC AROMATICS

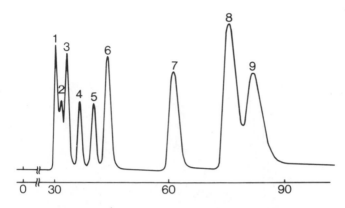

1. Benzene C_6H_6
2. Tetrahydronaphthalene $C_{10}H_{12}$
3. Styrene C_8H_8
4. Indene C_9H_8
5. Naphthalene $C_{10}H_8$

6. Biphenyl $C_{12}H_{10}$
7. Fluorene $C_{13}H_{10}$
8. Phenanthrene $C_{14}H_{10}$
9. Anthracene $C_{14}H_{10}$

Copper	2.00 m	2.00 mm
Alumina (Woelm)	100-125 μm	26.8 °C
Pentane		None
4 Bar	0.19 cm3/min	
Ultra-violet	254 nm	

A standard mixture; compare with chromatogram 1.08.

Martin, M., Loheac, J., Guiochon, G., *Chromatographia*, **5**, 33 (1972). Reproduced by permission of Friedr. Vieweg + Sohn GmbH, Braunschweig.

1.08 FRACTION FROM STEAM CRACKER, BOILING RANGE 150-350°C

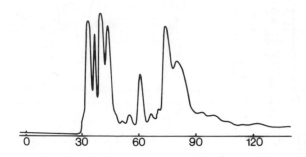

Copper	2.00 m	2.00 mm
Alumina (Woelm)	100-125 μm	26.8°C
Pentane		None
4 Bar	0.19 cm³/min	
Ultra-violet	254 nm	

See 1.07 for the chromatogram of standard compounds under identical conditions

Martin, M., Loheac, J., Guiochon, G., *Chromatographia,* **5**, 33 (1972). Reproduced by permission of Friedr. Vieweg + Sohn GmbH, Braunschweig.

1.09 POLYCYCLIC AROMATICS

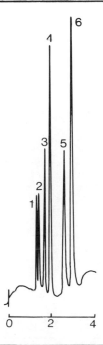

1. Toluene C_7H_8
2. Styrene C_8H_8
3. Naphthalene $C_{10}H_8$
4. Biphenyl $C_{12}H_{10}$
5. o-Terphenyl $C_{18}H_{14}$
6. Anthracene $C_{14}H_{10}$

Glass	0.22 m	5.5mm
Spherisorb (R) A 10 Y	17.5 μm	Ambient
Hexane saturated with water		None
14 Bar	0.28 cm/sec	1 μl
Ultra-violet	254 nm	0.08 AUFS

The very high efficiency may be due to the 'infinite diameter effect', — see page 12 in Part 1, since $Ld_p/d_c^2 = 0.12$

Knox, J.H., Laird, G., unpublished results

1.10 DODECYLBENZENE AND DIDODECYLBENZENE

1. Didodecylbenzene $C_{30}H_{54}$
2. Dodecylbenzene $C_{18}H_{30}$

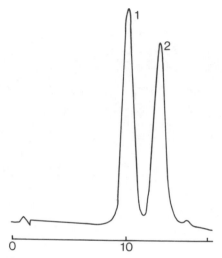

	4.00 m	
Poragel$^{(R)}$ A-3		
Tetrahydrofuran		None
	1.5 cm^3/min	
Refractive index		

An example of Exclusion Chromatography; the larger of the two molecules elutes first.

1.11 TERPHENYLS

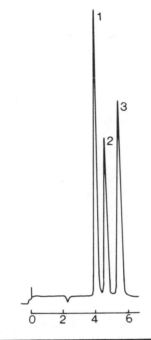

1. *o*-Terphenyl $C_{18}H_{14}$
2. *m*-Terphenyl $C_{18}H_{14}$
3. *p*-Terphenyl $C_{18}H_{14}$

Glass	0.50m	2.3 mm
Spherisorb$^{(R)}$ A 20 X	20 μm	Ambient
Hexane 50% saturated with water		None
22 Bar	0.38 cm/sec	4 μl
Ultra-violet	254 nm	0.16 AUFS

Partial deactivation of the alumina by the water reduced retention times and improved symmetry.

Done, J.N., Knox, J.H., unpublished results

2.01 FUMARIC AND MALEIC ACIDS

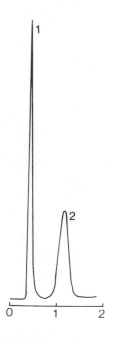

1. Fumaric acid $C_4H_4O_4$
2. Maleic acid $C_4H_4O_4$

Stainless steel	1.00 m	2.1 mm
Zipax (R) SAX	20-37 μm	60°C
0.01N Nitric acid		1% quaternary ammonium substituted polystyrene polymer
133 Bar	2.73 cm^3/min	3 μl
Ultra-violet	254 nm	

Kirkland, J.J., *Journal of Chromatographic Science,* **7,** 361 (1969). Reproduced by permission of Preston Technical Abstracts Co., Illinois.

2.02 IMPURITIES IN 3, 5, 6-TRICHLOROPYRIDIN-2-OL

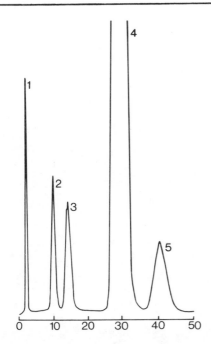

1. 2, 3, 5-Trichloro-6-ethoxypyridine $C_7H_6Cl_3NO$
2. 3, 5-Dichloropyridin-2-ol $C_5H_3Cl_2NO$
3. 3, 4, 5-Trichloropyridin-2-ol $C_5H_2Cl_3NO$
4. 3, 5, 6-Trichloropyridin-2-ol $C_5H_2Cl_3NO$
5. 3, 4, 5, 6-Tetrachloropyridin-2-ol C_5HCl_4NO

	0.50 m	2.8 mm
Bio-Rad (R) AGI-X2	37-74 μm	
Gradient: methanol to acetic acid at approximately 4% per minute		
	2.4 cm^3/min	10 μl
Ultra-violet	310 nm	0.2 AUFS

Gradient system was 'home made'. Due to large dead volume the gradient was not quite linear. Resin was in acetate form.

Skelly, N.E., Crummett, W.B., *Journal of Chromatography*, **55**, 309 (1971).
Reproduced by permission of Elsevier, Amsterdam.

2.03 PYRIDINECARBOXYLIC ACIDS

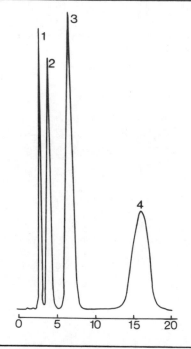

1. Pyridine-2-carboxylic acid $C_6H_5NO_2$
2. Pyridine-4-carboxylic acid $C_6H_5NO_2$
3. Pyridine-3-carboxylic acid $C_6H_5NO_2$
4. Picolinamide $C_6H_6N_2O$

Stainless steel	1.00 m	2.1 mm
Zipax$^{(R)}$ SCX	25-37 μm	27°C
Water containing 0.1N sodium nitrate and 0.1N phosphoric acid		1% sulfonated fluorocarbon
105 Bar	1.51 cm/sec	5 μl
Ultra-violet	254 nm	0.08 AUFS

Picolinamide used as an internal standard.

Talley, C.P., *Analytical Chemistry,* **43**, 1512 (1971). Copyright (1971) by the American Chemical Society. Reproduced by permission of the copyright owner.

2.04 AROMATIC CARBOXYLIC ACIDS

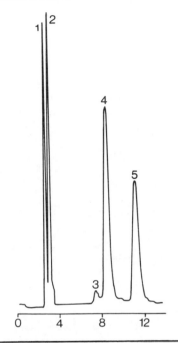

1. Acetone C_3H_6O
2. Benzoic acid $C_7H_6O_2$
3. Isophthalic acid $C_8H_6O_4$
4. Terephthalic acid $C_8H_6O_4$
5. Phthalic acid $C_8H_6O_4$

Stainless steel	1.00 m	2.1 mm
Zipax$^{(R)}$ AAX	37-44 μm	60°C
Water containing 0.01M citric acid adjusted to pH 3 with sodium hydroxide		1% quaternary ammonium substituted resin chemically bonded
21 Bar	0.66 cm/sec	2 μl
Ultra-violet	254 nm	0.01 AUFS

Knox, J.H., Vasvari, G., unpublished results

2.05 FREE FATTY ACIDS

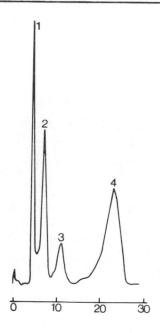

1. Hexanoic acid $C_6H_{12}O_2$
2. Dodecanoic acid $C_{12}H_{24}O_2$
3. Tetradecanoic acid $C_{14}H_{28}O_2$
4. Palmitic acid $C_{16}H_{32}O_2$

	1.00 m	6.0 mm
Silanized Celite	149-170 μm	Ambient
65% Acetone in water		20% n-octane
	0.28 cm/sec	25 μl
Transport detector		

The principle of the transport detector is described on page 11 of Part 1.
This separation was one of the first examples of the use of high speed liquid
chromatography.

Scott, R.P.W., Blackburn, D.W.J., Wilkins, T., *Journal of Gas Chromatography,* **5**, 183
(1967). Reproduced by permission of Preston Technical Abstracts Co., Illinois.

2.06 AROMATIC CARBOXYLIC ACIDS

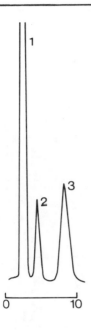

1. Benzoic acid $C_7H_6O_2$
2. *o*-Toluic acid $C_8H_8O_2$
3. Terephthalic acid $C_8H_6O_4$

Stainless steel	1.00 m	2.1 mm
Zipax$^{(R)}$ SAX	25-37 μm	
0.02M borate buffer with pH 9.2		1% quaternary ammonium substituted polystyrene polymer
84 Bar	1.5 cm^3/min	2 μl
Ultra-violet	254 nm	

Schmit, J.A., Henry, R.A., *Chromatographia*, **3**, 497 (1970). Reproduced by permission of Friedr. Vieweg + Sohn GmbH, Braunschweig.

2.07 TRICARBOXYLIC ACID CYCLE INTERMEDIATES

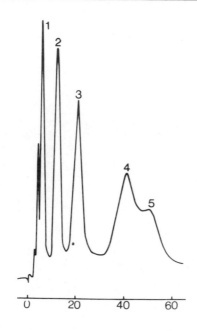

1. 2-Oxoglutaric acid $C_5H_6O_5$
2. *cis*-Propene-1,2,3-tricarboxylic acid $C_6H_6O_6$
3. Malic acid $C_4H_6O_5$
4. Citric acid $C_6H_8O_7$
5. Isocitric acid $C_6H_8O_7$

Stainless steel	0.30 m	2.3 mm
Silica gel	25-28 μm	19°C
Organic phase from mixture of 20 ml *t*-amyl alcohol, 50 ml chloroform and 50 ml 0.1N sulphuric acid.		30% of the aqueous phase from the same mixture
	0.47 cm³/min	1 μl
Reaction detector	432 nm	

Detector measured decrease in absorbance at 432 nm of a solution of the sodium salt of *o*-nitrophenol as the acids were eluted. The indicator and the eluent were mixed in a specially designed port.

Stahl, K.W., Schafer, G., Lamprecht, W., *Journal of Chromatographic Science,* **10**, 95 (1972). Reproduced by permission of Preston Technical Abstracts Co., Illinois.

2.08 TRICARBOXYLIC ACID CYCLE INTERMEDIATES

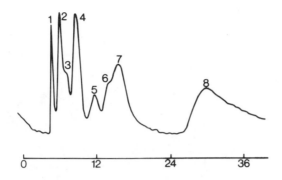

1. Acetic acid $C_2H_4O_2$
2. Fumaric acid $C_4H_4O_4$
3. Pyruvic acid $C_3H_4O_3$
4. Glutaric acid $C_5H_8O_4$

5. 3-Hydroxybutyric acid $C_4H_8O_3$
6. Lactic acid $C_3H_6O_3$
7. Succinic acid $C_4H_6O_4$
8. 2-Oxoglutaric acid $C_5H_6O_5$

Stainless steel	0.60 m	2.3 mm
Silica gel	15-25 μm	19°C
Organic phase from mixture of 6 ml t-amyl alcohol, 50 ml chloroform and 50 ml 0.01N sulphuric acid		40% of the aqueous phase from the same mixture
	0.75 cm^3/min	0.8 μl
Reaction detector	432 nm	

See Chromatogram 2.07 for details of the detector.

Stahl, K.W., Schafer, G., Lamprecht, W., *Journal of Chromatographic Science,* **10**, 95 (1972). Reproduced by permission of Preston Technical Abstracts Co., Illinois.

2.09 IMPURITIES IN 2,6-DIMETHYLPYRIDIN-4-OL

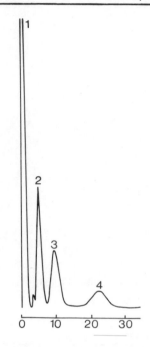

1. 2,6-Dimethylpyridin-4-ol C_7H_9NO
2. 6-Methylpyridine-2,4-diol $C_6H_7NO_2$
3. 4-Hydroxy-2,6-dimethylpyridine-3-carboxylic acid $C_8H_9NO_3$
4. Dehydroacetic $C_8H_8O_4$

	0.50 m	2.8 mm
Bio-Rad[(R)] AG1-X2	37-74 μm	
Gradient: methanol to 1% acetic acid in methanol		
	2.4 cm^3/min	10 μl
Ultra-violet	270 nm	0.2 AUFS

Gradient system was 'home made'. Time to complete gradient was about 25 minutes. Resin was in acetate form.

Skelly, N.E., Crummett, W.B., *Journal of Chromatography*, **55**, 309 (1971). Reproduced by permission of Elsevier, Amsterdam.

3.01 AROMATIC AMINES

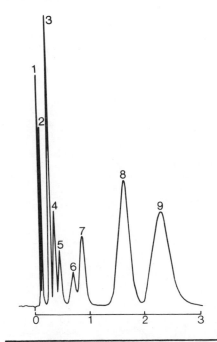

1. Azobenzene $C_{12}H_{10}N_2$
2. *N,N*-Dimethylaniline $C_8H_{11}N$
3. Benzo[*h*]quinoline $C_{13}H_9N$
4. Carbazole $C_{12}H_9N$
5. *o*-Toluidine C_7H_9N
6. 1-Naphthylamine $C_{10}H_9N$
7. 2-Naphthylamine $C_{10}H_9N$
8. Quinoline C_9H_7N
9. Isoquinoline C_9H_7N

Stainless steel	0.088 m	3.0 mm
Merckogel (R) SI 150	8-12 μm	Ambient
Iso-octane		None
130 Bar	6 cm^3/min	0.9 μl
Ultra-violet	270 nm	

Oster, H., Van Damme, S., Ecker, E., *Chromatographia*, **5**, 209 (1971). Reproduced by permission of Friedr. Vieweg + Sohn GmbH, Braunschweig.

3.02   DIAMINODIPHENYLMETHANES

1. Di-(*o*-aminophenyl)methane $C_{13}H_{14}N_2$
2. Di-(*p*-aminophenyl)methane $C_{13}H_{14}N_2$
3. Di-(*m*-aminophenyl)methane $C_{13}H_{14}N_2$

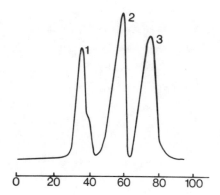

	0.15 m	6.0 mm
Aluminium oxide		18°C
1% ethanol in carbon tetrachloride		None
	0.44 cm³/min	

Kiselev, A.V., *Journal of Chromatography,* **49**, 84 (1970). Reproduced by permission of Elsevier, Amsterdam.

3.03 SIMPLE AROMATIC AMINES

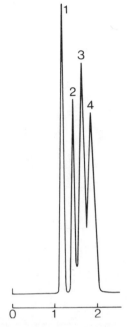

1. *m*-Xylene C_8H_{10}
2. 2,6-Dimethylaniline $C_8H_{11}N$
3. 2,3-Dimethylaniline $C_8H_{11}N$
4. Aniline C_6H_7N

Stainless steel	0.50 m	2.3 mm
Corasil$^{(R)}$ II	37-50 μm	Ambient
10% *iso*-propanol in hexane		None
	0.75 cm^3/min	
Refractive index		

Separation by adsorption

Little, J.N., Horgan, D.F., Bombaugh, K.J., *Journal of Chromatographic Science,* **8,** 625 (1970). Reproduced by permission of Preston Technical Abstracts Co., Illinois.

3.04 SIMPLE AROMATIC AMINES

1. Toluene C_7H_8
2. *N,N*-Dimethylaniline $C_8H_{11}N$
3. 2,6-Dimethylaniline $C_8H_{11}N$

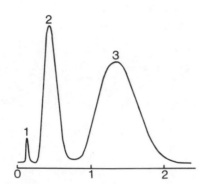

Steel	0.40 m	3.0 mm
Porasil$^{(R)}$ C	60-75 μm	20°C
Heptane		Propionitrile chemically bonded
29 Bar	5 cm/sec	
Ultra-violet		

Bonded stationary phase made by esterifying the Porasil with 3-hydroxyprop-ionitrile. This is one of the first examples of the use of a chemically bonded phase.

Halasz, I., Sebestian, I., *Angewandte Chemie International Edition in English,* **8**, 453 (1969). Reproduced by permission of Verlag Chemie GmbH, Weinheim.

3.05 SIMPLE AROMATIC AMINES

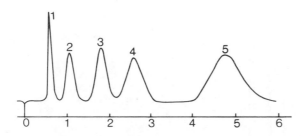

1. Benzene C_6H_6
2. Mesitylene C_9H_{12}
3. 2,6-Dimethylaniline $C_8H_{11}N$

4. 2,5-Dimethylaniline $C_8H_{11}N$
5. Aniline C_6H_7N

Steel	0.50 m	4.0 mm
Silica gel	50-71 μm	21°C
Heptane		10% PEG 400
13 Bar	1.34 cm/sec	
Ultra-violet		

Separation by partition.

Halasz, I., Gerlach, H.O., Kroneisen, A., Walkling, P., *Zeitschrift fur Analytische Chemie,* **234**, 97 (1968). Reproduced by permission of Springer-Verlag, Heidelberg.

3.06 AROMATIC AMINE ANTIOXIDANTS

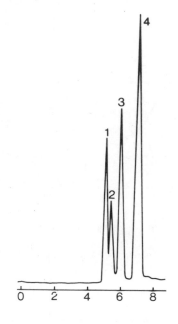

1. *N,N*-Diethylaniline $C_{10}H_{15}N$
2. *N*-Ethylaniline $C_8H_{11}N$
3. Diphenylamine $C_{12}H_{11}N$
4. *N*-Phenyl-2-naphthylamine $C_{16}H_{13}N$

Stainless steel	1.00 m	2.1 mm
Zipax(R)	20-37 μm	Ambient
Iso-octane		0.5% BOP
	0.31 cm³/min	10.6 μl
Ultra-violet	254 nm	

A six-port, rotary liquid sampling valve was used for injection.

Majors, R.E., *Journal of Chromatographic Science,* **8**, 338 (1970). Reproduced by permission of Preston Technical Abstracts Co., Illinois.

3.07 PHENYLENEDIAMINES

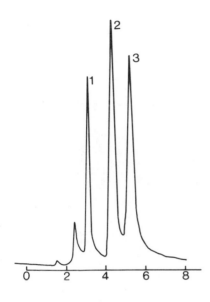

1. *o*-Phenylenediamine $C_6H_8N_2$
2. *m*-Phenylenediamine $C_6H_8N_2$
3. *p*-Phenylenediamine $C_6H_8N_2$

Stainless steel	1.00 m	2.1 mm
Permaphase(R)ETH	<37μm	27°C
5% methanol in cyclopentane		Ether chemically bonded
18 Bar	1.00 cm³/min	6 μl
Ultra-violet	254 nm	

Column prepared using a high pressure slurry-packing procedure.

Kirkland, J.J., *Journal of Chromatographic Science*, **9**, 206 (1971). Reproduced by permission of Preston Technical Abstracts Co., Illinois.

3.08 NITROANILINES

1. *o*-Nitroaniline $C_6H_6N_2O_2$
2. *m*-Nitroaniline $C_6H_6N_2O_2$
3. *p*-Nitroaniline $C_6H_6N_2O_2$

Stainless steel	1.50 m	2.3 mm
Durapak (R)	36-75 μm	Ambient
Chloroform		Polyethylene glycol chemically bonded
25 Bar	1.75 cm³/min	
Refractive index		

Bombaugh, K.J., Levangie, R.F., King, R.N., Abrahams, L., *Journal of Chromatographic Science,* **8**, 657 (1970). Reproduced by permission of Preston Technical Abstracts Co., Illinois.

3.09 AROMATIC BASES

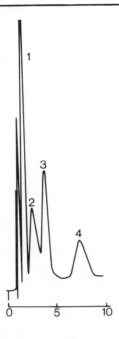

1. Pyridine C_5H_5N
2. 8-Quinolinol C_9H_7NO
3. Quinoline C_9H_7N
4. Isoquinoline C_9H_7N

Stainless steel	1.00 m	2.1 mm
Zipax$^{(R)}$SCX	25-37 μm	
Water containing 0.15M sodium nitrate		1% sulfonated fluorocarbon
84 Bar		
Ultra-violet	254 nm	

Chromatographic Methods 820M5. Reproduced by permission of Du Pont Instrument Products Division, Wilmington.

3.10 AZA-HETEROCYCLICS

1. Benzo[h]quinoline $C_{13}H_9N$
2. Phenazine $C_{12}H_8N_2$
3. Acridine $C_{13}H_9N$
4. Benzo[c]quinoline $C_{13}H_9N$
5. Benzo[f]quinoline $C_{13}H_9N$

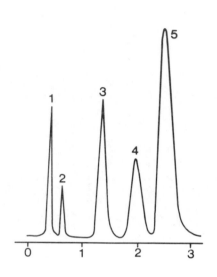

Stainless steel	1.00 m	2.4 mm
Zipax[R]	37-44 μm	23°C
1% acetonitrile in hexane		0.3% silver ion
62 Bar	4 cm/sec	
Ultra-violet	254/280 nm	

Stationary phase prepared by reacting sodium hydroxide with silver nitrate on the Zipax [R].

Vivilecchia, R., Thiebaud, M., Frei, R.W., *Journal of Chromatographic Science,* **10***,* 411 (1972). Reproduced by permission of Preston Technical Abstracts Co., Illinois.

3.11 BENZOQUINOLINES

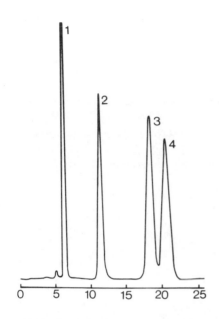

1. Benzo[h] quinoline $C_{13}H_9N$
2. Acridine $C_{13}H_9N$
3. Benzo[c] quinoline $C_{13}H_9N$
4. Benzo[f] quinoline $C_{13}H_9N$

Glass	0.50 m	2.1 mm
Spherisorb[(R)] A 20 Y	20 μm	Ambient
Hexane		5% BOP
21 Bar		1.25 μl
Ultra-violet	254 nm	0.08 AUFS

This separation may in fact be due to adsorption and not partition; the BOP merely deactivating the alumina whereas water is normally used for this purpose.

Done, J.N., Knox, J.H., unpublished results

3.12 AZOBENZENE AND AMINOAZOBENZENE DERIVATIVES

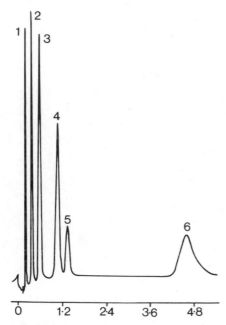

1. Azobenzene $C_{12}H_{10}N_2$
2. 3'-Bromo-4-diethylaminoazobenzene $C_{16}H_{18}BrN_3$
3. 4-Diethylaminoazobenzene $C_{16}H_{19}N_3$
4. 4-Diethylamino-3'-nitroazobenzene $C_{16}H_{18}N_4O_2$
5. 4-Diethylamino-4'-nitroazobenzene $C_{16}H_{18}N_4O_2$
6. 4-Aminoazobenzene $C_{12}H_{11}N_3$

Stainless steel	0.15 m	2.4 mm
Lichrosorb(R) SI 60	10 μm	Ambient
10% methylene chloride in hexane		None
24 Bar	2.2 cm³/min	1 μl
Ultra-violet	254 nm	0.12 AUFS

An example of a column packed using a balanced density slurry.

3.13 AMIDOPYRIDINES

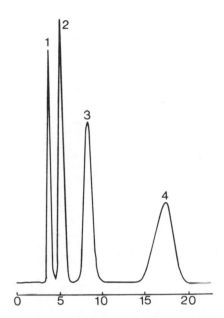

1. Nicotinonitrile $C_6H_4N_2$
2. Isonicotinamide $C_6H_6N_2O$
3. Nicotinamide $C_6H_6N_2O$
4. Picolinamide $C_6H_6N_2O$

Stainless steel	1.00 m	2.1 ιιιιι
Zipax(R) SCX	25-37 μm	27°C
Water containing 0.1N sodium nitrate and 0.1N phosphoric acid		1% sulfonated fluorocarbon
105 Bar	1.51 cm/sec	5 μl
Ultra-violet	254 nm	0.08 AUFS

Nicotinonitrile used as internal standard.

3.14 CYANOPYRIDINES

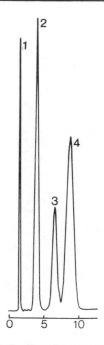

1. Picolinonitrile $C_6H_4N_2$
2. Nicotinonitrile $C_6H_4N_2$
3. Isonicotinonitrile $C_6H_4N_2$
4. Nicotinamide $C_6H_6N_2O$

Stainless steel	1.00 m	2.1 mm
Zipax$^{(R)}$ SCX	25-37 μm	27°C
Water containing 0.1N sodium nitrate and 0.1N phosphoric acid		1% sulfonated fluorocarbon
105 Bar	1.51 cm/sec	5 μl
Ultra-violet	254 nm	0.08 & 0.04 AUFS

Nicotinamide used as internal standard. Five analyses gave standard deviations of 1.35%, 0.73% and 0.77% for the three nitriles.

4.01 IMPURITIES IN 2,2',6,6'-TETRABROMOBISPHENOL A

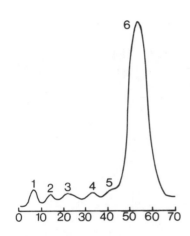

1. Bisphenol A $C_{15}H_{16}O_2$
2. 2-Bromobisphenol A $C_{15}H_{15}BrO_2$
3. 2,2'-Dibromobisphenol A
   $C_{15}H_{14}Br_2O_2$
4. 2,6-Dibromobisphenol A
   $C_{15}H_{14}Br_2O_2$
5. 2,2',6-Tribromobisphenol A
   $C_{15}H_{13}Br_3O_2$
6. 2,2',6,6'-Tetrabromobisphenol A
   $C_{15}H_{12}Br_4O_2$

	0.50 m	2.8 mm
Bio-Rad$^{(R)}$ AG1-X2	37-74 μm	Ambient
Gradient: methanol to 1% acetic acid in methanol		
	2.4 cm^3/min	5 μl
Ultra-violet	285 nm	0.2 AUFS

Gradient system was 'home made'. Time to complete gradient was about 60 minutes Resin was in acetate form.

Skelly, N.E., Crummett, W.B., *Journal of Chromatography,* **55**, 309 (1971). Reproduced by permission of Elsevier, Amsterdam.

4.02 PHENOLS

1. 2,6-Xylenol $C_8H_{10}O$
2. 2,3-Xylenol $C_8H_{10}O$
3. 3,4-Xylenol $C_8H_{10}O$
4. 3,5-Xylenol $C_8H_{10}O$
5. p-Cresol C_7H_8O
6. m-Cresol C_7H_8O
7. Phenol C_6H_6O

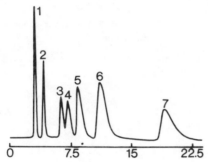

Stainless steel	1.00 m	2.1 mm
Permaphase[R] ETH	37 μm	
2.5% methanol in cyclopentane		0.88% ether chemically bonded
17.5 Bar	1 cm^3/min	20 μl
Ultra-violet	254 nm	

Kirkland, J.J., *Journal of Chromatographic Science,* **9**, 206 (1971). Reproduced by permission of Preston Technical Abstracts Co., Illinois.

4.03 HINDERED PHENOLS

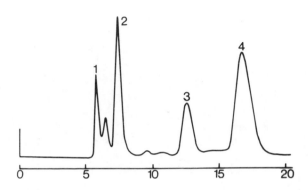

1. 2,4,6-Tri-t-butylphenol $C_{18}H_{30}O$
2. 6-t-Butyl-o-cresol $C_{11}H_{16}O$

3. 2,6-Xylenol $C_8H_{10}O$
4. 2-t-Butylphenol $C_{10}H_{14}O$

Glass	0.50 m	2.0 mm
Corasil(R) II	44-53 μm	Ambient
Cyclohexane		None
	0.15 cm/sec	5 μl
Ultra-violet	254 nm	

Done, J.N., Kennedy, G.J., Knox, J.H., *Nature,* **237**, 77 (1972). Reproduced by permission of MacMillan (Journals) Ltd., London.

4.04 PHENOLS

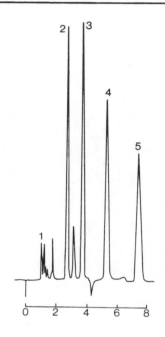

1. Unretained solute
2. 2,4-Xylenol $C_8H_{10}O$
3. o-Cresol C_7H_8O
4. m-Cresol C_7H_8O
5. Phenol C_6H_6O

Glass	1.00 m	2.0 mm
Zipax[(R)]	29 μm	Ambient
Hexane		2% BOP
35 Bar	1.66 cm/sec	1 μl
Ultra-violet	254 nm	0.16 AUFS

This compares very well with analysis of similar mixtures by GC.

Done, J.N., Kennedy, G.J., Knox, J.H., *Nature,* **237**, 77 (1972). Reproduced by permission of MacMillan (Journals) Ltd., London.

4.05 IMPURITIES IN 2,4,6-TRICHLOROPHENOL

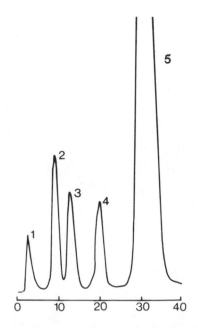

1. p-Chlorophenol C_6H_5ClO
2. o-Chlorophenol C_6H_5ClO
3. 2,4-Dichlorophenol $C_6H_4Cl_2O$
4. 2,6-Dichlorophenol $C_6H_4Cl_2O$
5. 2,4,6-Trichlorophenol $C_6H_3Cl_3O$

	0.50 m		2.8 mm
BioRad$^{(R)}$ AG1-X2	37-74 μm		
Gradient: methanol to 5% acetic acid in methanol			
	2.4 cm^3/min		10 μl
Ultra-violet	254 nm		0.2 AUFS

Gradient system was 'home made'. Time to complete gradient was about 60 minutes.

Skelly, N.E., Crummett, W.B., *Journal of Chromatography*, **55**, 309 (1971). Reproduced by permission of Elsevier, Amsterdam.

4.06 PHENOLS

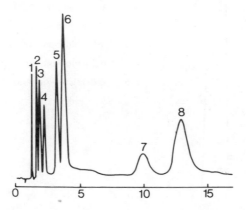

1. o-Nitrophenol $C_6H_5NO_3$
2. 1-Naphthol $C_{10}H_8O$
3. 2-Naphthol $C_{10}H_8O$
4. m-Nitrophenol $C_6H_5NO_3$

5. p-Nitrophenol $C_6H_5NO_3$
6. Pyrocatechol $C_6H_6O_2$
7. Resorcinol $C_6H_6O_2$
8. Hydroquinone $C_6H_6O_2$

	0.20 m		3.0 mm
Chromosorb(R) G			20°C
20% ether in iso-octane		5% Fractonitril III	
	1.2 cm^3/min		
Ultra-violet	280 nm		

Siemens' liquid chromatography bulletin S 200. Reproduced by permission of Siemens AG, Karlsruhe.

4.07 SUBSTITUTED DIHYDROXYBENZENES

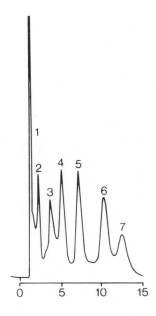

1. 2,5-Di-t-butylhydroquinone $C_{14}H_{22}O_2$
2. t-Butylhydroquinone $C_{10}H_{14}O_2$
3. Pyrocatechol $C_6H_6O_2$
4. Bromohydroquinone $C_6H_5BrO_2$
5. Naphthalene-1,5-diol $C_{10}H_8O_2$
6. Resorcinol $C_6H_6O_2$
7. Hydroquinone $C_6H_6O_2$

Stainless steel	1.00 m	2.1 mm
Zipax$^{(R)}$	37-44 μm	Ambient
10% tetrahydrofuran in heptane		1% BOP
72 Bar	1 cm^3/min	
Ultra-violet	254 nm	

Schmit, J.A., 'Applications of High Speed Liquid Chromatography Using CSP Supports' in *Modern Practice of Liquid Chromatography* edited by Kirkland, J.J. Copyright (C), John Wiley and Sons, Inc., (1971). Reproduced by permission of John Wiley and Sons, Inc.

4.08 t-BUTYLPHENOLS

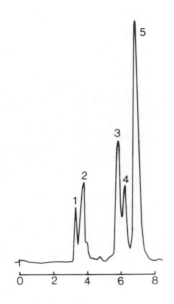

1. 2,4,6-Tri-t-butylphenol $C_{18}H_{30}O$
2. 2-t-Butyl-4,6-dimethylphenol $C_{12}H_{18}O$
3. 6-t-Butyl-*m*-cresol $C_{11}H_{16}O$
4. 6-t-Butyl-*p*-cresol $C_{11}H_{16}O$
5. 2-t-Butylphenol $C_{10}H_{14}O$

Stainless steel	1.00 m	2.1 mm
Corasil$^{(R)}$ II	37-44 μm	Ambient
12.5% chloroform in hexane		None
7 bar	0.51 cm/sec	0.8 μl
Ultra-violet	254 nm	0.08 AUFS

Done, J.N., Knox, J.H., unpublished results

4.09   ANTIOXIDANTS

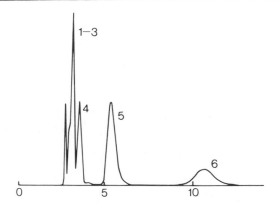

1. 2,6-Di-t-butyl-p-cresol $C_{15}H_{24}O$

2. 4,4'-Methylenebis(2,6-di-t-butylphenol) $C_{29}H_{44}O_2$

3. 1,3,5-Tris(3,5-di-t-butyl-4-hydroxyphenyl)-2,4,6-trimethyl benzene $C_{51}H_{72}O_3$

4. 4,4'-Methylenebis(6-t-butyl-o-cresol) $C_{23}H_{32}O_2$

5. 4,4'-Thiobis(6-t-butyl-o-cresol) $C_{22}H_{30}O_2S$

6. 2-Hydroxy-4-methoxybenzophenone $C_{14}H_{12}O_3$

	1.50 m	
Corasil(R) II	36-75 μm	Ambient
17% methylene chloride in pentane		None
Ultra-violet	285 nm	

Corasil(R) II does not resolve the first three components but by using a coupled column system with added Porasil(R) A columns all six components were resolved.

Snyder, L.R., 'The Practice of Liquid-Solid Chromatography' in *Modern Practice of Liquid Chromatography* edited by Kirkland, J.J. Copyright (C), John Wiley and Sons, Inc., (1971). Reproduced by permission of John Wiley and Sons, Inc.

4.10 COMMERCIAL ANTIOXIDANTS

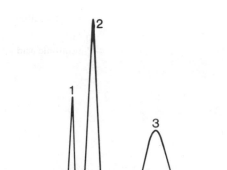

1. Irganox 1076 $C_{35}H_{62}O_3$
2. CAO-14
3. Santonox R $C_{22}H_{30}O_2S_2$

Stainless steel	1.0 m	2.1 mm
Corasil$^{(R)}$ II	37-50 μm	Ambient
1% isopropanol in hexane		None
	0.95 cm^3/min	10.6 μl
Ultra-violet	254 nm	

A six-port, rotary liquid sampling valve was used for injection.

Majors, R.E., *Journal of Chromatographic Science,* **8**, 338 (1970). Reproduced by permission of Preston Technical Abstracts Co., Illinois.

5.01 BIPHENYLSULFONIC ACIDS

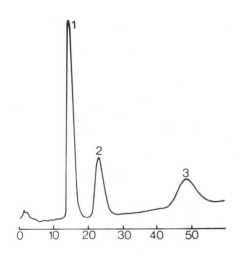

1. Biphenyl-4-sulfonic acid
    $C_{12}H_{10}O_3S$
2. 4'-Hydroxybiphenyl-4-sulfonic
    acid $C_{12}H_{10}O_4S$
3. Biphenyl-4-4'-disulfonic acid
    $C_{12}H_{10}O_6S_2$

Glass	0.50 m	2 mm
Bio-Rad[R] Bio-Rex 5	44-52 μm	Ambient
Gradient: water:acetonitrile:methanol in the ratio 1:1:1 to 1.0M LiCl in the same solvent		
	0.5 cm³/min	10 μl
Ultra-violet	250 nm	0.4 AUFS

Gradient system was 'home made'. LiCl solution added with mixing to 40 ml of solvent at the same rate as the mobile phase is withdrawn from the mixing vessel.

Stehl, R.H., *Analytical Chemistry*, **42**, 1802 (1970). Copyright(1970)by the American Chemical Society. Reproduced by permission of the copyright owner.

5.02 AROMATIC SULFONIC ACIDS

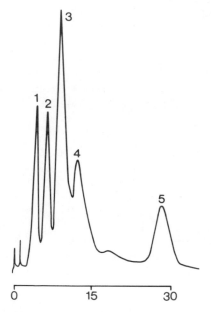

1. Benzenesulfonic acid $C_6H_6O_3S$
2. Toluene-*p*-sulfonic acid $C_7H_8O_3S$
3. 2,5-Dimethylbenzenesulfonic acid $C_8H_{10}O_3S$
4. *p*-Chlorobenzenesulfonic acid $C_6H_5ClO_3S$
5. Naphthalene-2-sulfonic acid $C_{10}H_8O_3S$

Stainless steel	1.00 m	2.1 mm
Zipax[(R)] SAX	25-37 μm	60°C
Gradient: water containing 0.0025M perchloric acid to water containing 0.005M perchloric acid. Stepwise after 18 minute		1% quaternary ammonium substituted polystyrene polymer
26 Bar	0.54 cm^3/min	
Ultra-violet	254 nm	0.1 AUFS

Kirkland, J.J., *Analytical Chemistry*, **43**, No. 12, 41A (1971). Copyright(1971) by the American Chemical Society. Reproduced by permission of the copyright owner.

5.03 AMINOSULFONIC ACIDS

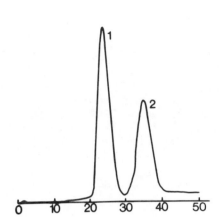

1. 4,6-Dimethylorthanilic acid
   $C_8H_{11}NO_3S$
2. Orthanilic acid $C_6H_7NO_3S$

Glass	0.50 m	2.0 mm
Bio-Rad(R) Bio-Rex 5	44-52 μm	Ambient
Gradient: water:acetonitrile:methanol in the ratio 1:1:1 to 0.66M LiCl in the same solvent		
	1.0 cm^3/min	10 μl
Ultra-violet	250 nm	0.4 AUFS

Gradient system was 'home made'. LiCl solution added with mixing to 50 ml of solvent at the same rate as the mobile phase is withdrawn from the mixing vessel

Stehl, R.H. *Analytical Chemistry,* **42**, 1802 (1970). Copyright (1970) by the American Chemical Society. Reproduced by permission of the copyright owner.

5.04 DYE INTERMEDIATES

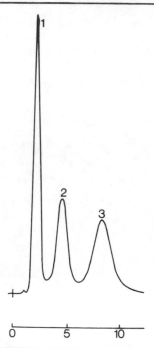

1. Schaeffer's salt $C_{10}H_8O_4S$
2. R salt $C_{10}H_8O_7S_2$
3. G salt $C_{10}H_8O_7S_2$

Stainless steel	1.0 m	2.1 mm
Zipax$^{(R)}$ SAX	25-37 μm	Ambient
Water containing 0.025M sodium nitrate		1% quaternary ammonium substituted polystyrene polymer
84 Bar	1.5 cm^3/min	2 μl
Ultra-violet	254 nm	

Schmit, J.A., Henry, R.A., *Chromatographia,* **3**, 497 (1970). Reproduced by permission of R. Henry.

5.05 IMPURITIES IN AMINOANTHRAQUINONE DYE

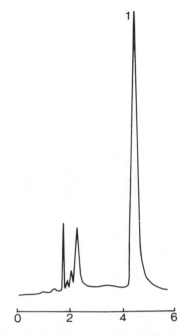

1. N-Benzoyl-1,5-diaminoanthraquinone
   $C_{21}H_{14}N_2O_3$

Stainless steel	1.0 m	2.1 ꞁꞁꞁꞁꞁ
Zipax$^{(R)}$	37-44 μm	Ambient
10% tetrahydrofuran in heptane		1% BOP
42 Bar	1 cm^3/min	
Ultra-violet	254 nm	

Schmit, J.A., 'Applications of High Speed Liquid Chromatography Using CSP Supports' in *Modern Practice of Liquid Chromatography* edited by Kirkland, J.J. Copyright (C), John Wiley and Sons, Inc., (1971). Reproduced by permission of John Wiley and Sons, Inc.

5.06 FOOD COLORING AND PRODUCTION INTERMEDIATES

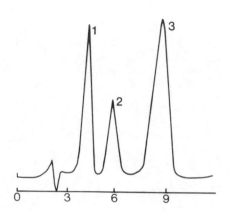

1. 3-Carboxy-5-hydroxy-1-*p*-sulfophenylpyrazole $C_{10}H_8N_2O_6S$
2. *p*-Anilinesulfonic acid $C_6H_7NO_3S$
3. FD&C Yellow 5 $C_{16}H_{12}N_4O_9S_2$

	0.50 m	3 mm
Sil-X$^{(R)}$	36-45 μm	Ambient
Gradient: 16% tetrahydrofuran in methanol to 33.3% tetrahydrofuran (containing 1% acetic acid) in methanol.		None
17 Bar	1.6 cm^3/min	
Ultra-violet		0.1 AUFS

Reproduced by courtesy of Perkin-Elmer Inc.

5.07 HYDROXYBENZENESULFONIC ACIDS

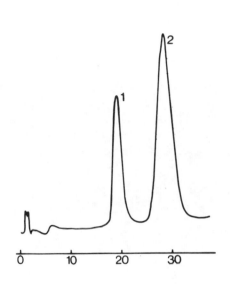

1. Benzenesulfonic acid $C_6H_6O_3S$
2. p-Hydroxybenzenesulfonic acid
   $C_6H_6O_4S$

Glass	0.50 m	2 mm
Bio-Rad$^{(R)}$ Bio-Rex 5	44-52 μm	Ambient
Gradient: water:acetonitrile:methanol in the ratio 1:1:1 to 1.0M LiCl in the same solvent		
	1 cm^3/min	10 μl
Ultra-violet	250 nm	0.4 AUFS

Gradient system was 'home made'. LiCl solution added with mixing to 50 ml of solvent at the same rate as the mobile phase is withdrawn from the mixing vessel. o-Hydroxybenzenesulfonic acid under the same conditions elutes between peaks 1 and 2.

6.01 SUBSTITUTED UREA HERBICIDES

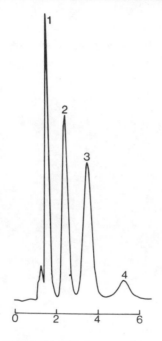

1. Linuron $C_9H_{10}Cl_2N_2O_2$
2. Diuron $C_9H_{10}Cl_2N_2O$
3. Monuron $C_9H_{11}ClN_2O$
4. Fenuron $C_9H_{12}N_2O$

Stainless steel	0.50 m	2.1 mm
Zipax(R)	37-44 μm	Ambient
Dibutyl ether		1% BOP
	1.14 cm^3/min	1 μl
Ultra-violet	254 nm	

First example of the use of Zipax(R) to give rapid efficient separation of polar compounds.

Kirkland, J.J., *Journal of Chromatographic Science,* **7**, 7 (1969). Reproduced by permission of Preston Technical Abstracts Co., Illinois.

6.02 INSECTICIDES

1. Parathion $C_{10}H_{14}NO_5PS$
2. Folpet $C_9H_4Cl_3NO_2S$
3. Imidan $C_{11}H_{12}NO_4PS_2$

Glass	0.50 m	2 mm
Permaphase$^{(R)}$ ETH	37-44 μm	Ambient
Iso-octane		1% ether chemically bonded
6.2 Bar	0.4 cm^3/min	40 μl
Ultra-violet	254 nm	0.08 AUFS

6.03 INSECTICIDES

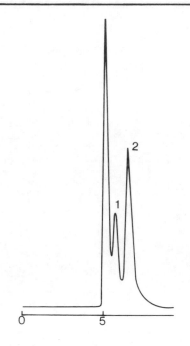

1. Parathion $C_{10}H_{14}NO_5PS$
2. Methylparathion $C_8H_{10}NO_5PS$

Steel	1.0 m	2.0 mm
Alumina type T	30-40 μm	Ambient
40% benzene in diethyl ether		None
	0.33 cm/sec	
Refractive index		

Separation by adsorption.

Randau, D., Bayer, H., *Journal of Chromatography*, **66**, 382 (1972). Reproduced by permission of Elsevier, Amsterdam.

6.04 TECHNICAL INSECTICIDE MIXTURE

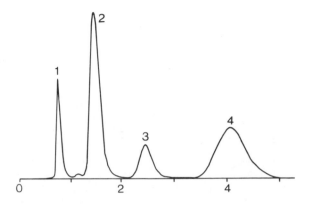

1. *p*-Nitrophenol $C_6H_6NO_3$
2. Methylparathion $C_8H_{10}NO_5PS$

3. Oxygen O_2
4. Parathion $C_{10}H_{14}NO_5PS$

Glass	0.20 m	2.75 mm
Diatomaceous earth	28-40 μm	Ambient
Water containing 38.8% ethanol, 0.80% acetic acid, 0.21% sodium hydroxide and 0.09% potassium chloride		10% iso-octane
	1 cm^3/min	10 μl
Polarograph		

The polarograph is a very sensitive detector but has limited use due to its specificity. Support was silanised.

Koen, J.G., Huber, J.F.K., Poppe, H., den Boef, G., *Journal of Chromatographic Science,* **8**, 192 (1970). Reproduced by permission of Preston Technical Abstracts Co., Illinois.

6.05 IMPURITIES IN ABATE^(R)

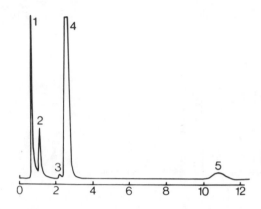

1. Impurity
2. Impurity
3. Impurity

4. Abate^(R) $C_{16}H_{20}O_6P_2S_3$
5. Impurity

Stainless steel	1.0 m	2.1 mm
Zipax^(R)	37-44 μm	Ambient
Heptane		1% BOP
Ultra-violet	254 nm	

Henry, R.A., Schmit, J.A., Dieckman, J.F., Murphey, F.J., *Analytical Chemistry,* **43**, 1053 (1971). Copyright (1971) by the American Chemical Society. Reproduced by permission of the copyright owner.

6.06 PESTICIDE RESIDUE ANALYSIS

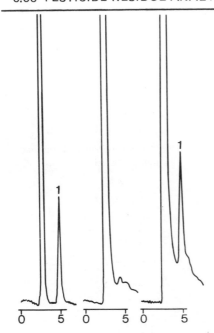

1. Dyfonate$^{(R)}$ $C_{10}H_{15}O_2PS_2$

Glass	1.0 m	2 mm
Corasil$^{(R)}$ II	37-50 μm	Ambient
2.5% chloroform in iso-octane		None
16.5 Bar	1 cm^3/min	40 μl
Ultra-violet	254 nm	0.02 AUFS

Chromatograms show, from left to right, a 1 ppm pesticide standard, the crop extract, and this extract fortified with the pesticide standard. No cleanup was necessary. The alfalfa merely being extracted with benzene and the extract concentrated.

Chromatronix Inc., Liquid Chromatography application Number 9. Reproduced by permission of Chromatronix Inc., Berkeley.

6.07   2,4-DICHLOROPHENOXYACETIC ACID AND ESTERS

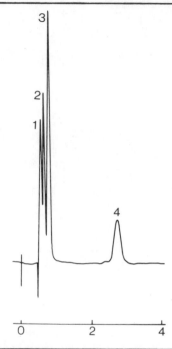

1. 2,4-Dichlorophenoxyacetic acid $C_8H_6Cl_2O_3$
2. 2,4-Dichlorophenoxyacetic acid, isopropyl ester $C_{11}H_{12}Cl_2O_3$
3. 2,4-Dichlorophenoxyacetic acid, isobutyl ester $C_{12}H_{14}Cl_2O_3$
4. 2,4-Dichlorophenoxyacetic acid, ethylhexyl ester $C_{16}H_{22}Cl_2O_3$

Stainless steel	1.0 m	2.1 mm
Permaphase(R) ODS	37-44 μm	60°C
40% water in methanol		1% octadecylsilyl chemically bonded
84 Bar	2.2 cm³/min	
Ultra-violet	254 nm	

Schmit, J.A., 'Applications of High Speed Liquid Chromatography Using CSP Supports' in *Modern Practice of Liquid Chromatography* edited by Kirkland, J.J. Copyright (C), John Wiley and Sons, Inc., (1971). Reproduced by permission of John Wiley and Sons, Inc.

6.08 CARBARYL IN PLANT EXTRACT

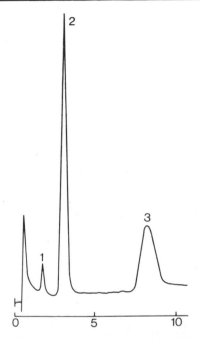

1. Impurity
2. Carbaryl $C_{12}H_{11}NO_2$
3. 1-Naphthol $C_{10}H_8O$

Stainless steel	1.0 m	2.1 mm
Zipax$^{(R)}$	37-44 μm	Ambient
Hexane		1% trimethylene glycol
84 Bar	2 cm^3/min	
Ultra-violet	254 nm	

1-Naphthol is the hydrolysis product of carbaryl.

6.09 INSECTICIDE

1. Methoxychlor $C_{16}H_{15}Cl_3O_2$
2. 1,1-Dichloro-2,2-di(p-methoxy phenyl)ethane $C_{16}H_{16}Cl_2O_2$

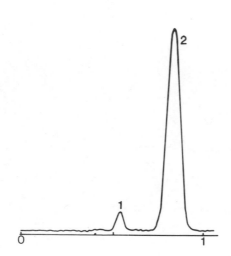

Glass	0.50 m	3.2 mm
Zipax$^{(R)}$	37-44 μm	27°C
Hexane		0.5% BOP
8 Bar	2.01 cm/sec	2 μl
Ultra-violet	254 nm	0.05 AUFS

6.10 CHLORINATED INSECTICIDES

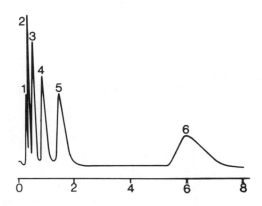

1. Impurity
2. Aldrin $C_{12}H_8Cl_6$
3. p,p'-D.D.T. $C_{14}H_9Cl_5$

4. D.D.D. $C_{14}H_{10}Cl_4$
5. Lindane[R] $C_6H_6Cl_6$
6. Endrin $C_{12}H_8Cl_6O$

	0.50 m	2.3 mm
Corasil[R] II	37-50 μm	Ambient
10% isopropanol in hexane		None
	0.75 cm³/min	
Refractive index		

Little, J.N., Horgan, D.F., Bombaugh, K.J., *Journal of Chromatographic Science,* **8,** 625 (1970). Reproduced by permission of Preston Technical Abstracts Co., Illinois.

6.11 TANDEX$^{(R)}$ IN SOIL EXTRACT

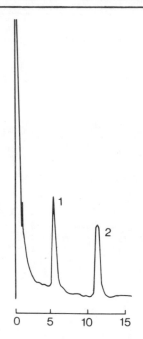

1. Tandex$^{(R)}$ $C_{14}H_{21}N_3O_3$
2. Hydrolysis product

Stainless steel	1.0 m	2.1 mm
Zipax$^{(R)}$	37-44 μm	Ambient
5% tetrahydrofuran in heptane		1% BOP
59 Bar	1.4 cm^3/min	
Ultra-violet	254 nm	

Schmit, J.A., 'Applications of High Speed Liquid Chromatography Using CSP Supports' in *Modern Practice of Liquid Chromatography* edited by Kirkland, J.J. Copyright (C), John Wiley and Sons, Inc., (1971). Reproduced by permission of John Wiley and Sons, Inc.

6.12 THIOLHYDROXAMATES

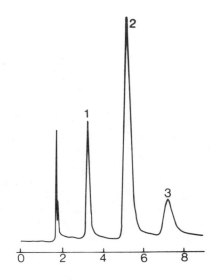

1. *S*-Methyl (*N*-methylcarbamoyloxy) thioacetimidate $C_5H_{10}N_2O_2S$

2. *S*-Methyl (dimethylcarbamoyl)-*N*-(methylcarbamoyloxy)thiocarbamate $C_7H_{13}N_3O_3S$

3. *S*-Methyl *N*-(carbamoyloxy) thio acetimidate $C_4H_8N_2O_2S$

Stainless steel	1.0 m	2.1 mm
Permaphase$^{(R)}$ ETH	<37 μm	60°C
10% dioxan in iso-octane		0.88% ether chemically bonded
15 Bar	0.86 cm^3/min	
Ultra-violet	254 nm	

Column prepared using a high pressure slurry-packing procedure.

Kirkland, J.J., *Journal of Chromatographic Science,* **9**, 206 (1971). Reproduced by permission of Preston Technical Abstracts Co., Illinois.

6.13 LANNATE[(R)] INSECTICIDE

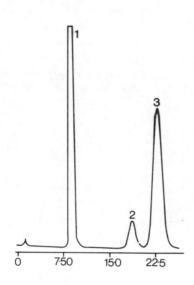

1. Benzanilide $C_{13}H_{11}NO$
2. *S*-Methyl *N*-hydroxythioacetimidate C_3H_7NOS
3. Methomyl $C_5H_{10}N_2O_2S$

Stainless steel	1.0 m	2.1 mm
Zipax[(R)]	20-37 μm	Ambient
7% chloroform in hexane		1% BOP
	1.3 cm^3/min	20 μl
Ultra-violet	254 nm	0.08 AUFS

Benzanilide used as internal standard

Leitch, R.E., *Journal of Chromatographic Science,* **9**, 531 (1971). Reproduced by permission of Preston Technical Abstracts Co., Illinois.

6.14 INSECTICIDES

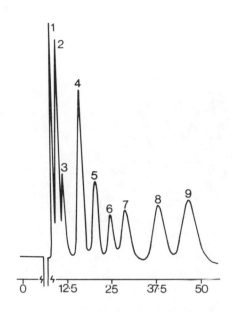

1. Methylparathion $C_8H_{10}NO_5PS$
2. Aldrin $C_{12}H_8Cl_6$
3. Heptachlor $C_{10}H_5Cl_7$
4. o,p'-D.D.T. $C_{14}H_9Cl_5$
5. p,p'-D.D.T. $C_{14}H_9Cl_5$
6. O,O-Dimethyl chlorothiophosphate $C_2H_6ClO_2PS$
7. p,p'-D.D.D. $C_{14}H_{10}Cl_4$
8. Lindane[R] $C_6H_6Cl_6$
9. Endrin $C_{12}H_8Cl_6O$

	1.20 m	2 mm
Porasil[R] 60	37-75 μm	Ambient
Iso-octane		10% BOP
	0.80 cm^3/min	
Refractive index		

The negative peak at the start of the chromatogram is EPN[R] $C_{14}H_{15}NO_4PS$

Waters, J.L., Little, J.N., Horgan, D.F., *Journal of Chromatographic Science,* **7**, 294 (1969).
Reproduced by permission of Preston Technical Abstracts Co., Illinois.

7.01   RIBONUCLEOSIDE MONO-, DI-, AND TRIPHOSPHORIC ACIDS

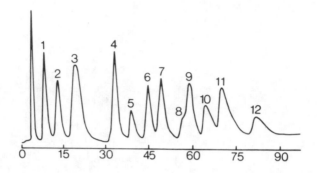

1. 5'-CMP $C_9H_{14}N_3O_8P$
2. 5'-UMP $C_9H_{13}N_2O_9P$
3. 5'-AMP $C_{10}H_{14}N_5O_7P$
4. 5'-GMP $C_{10}H_{14}N_5O_8P$
5. 5'-CDP $C_9H_{15}N_3O_{11}P_2$
6. 5'-UDP $C_9H_{14}N_2O_{12}P_2$
7. 5'-ADP $C_{10}H_{15}N_5O_{10}P_2$
8. 5'-CTP $C_9H_{16}N_3O_{14}P_3$
9. 5'-GDP $C_{10}H_{15}N_5O_{11}P_2$
10. 5'-UTP $C_9H_{15}N_2O_{15}P_3$
11. 5'-ATP $C_{10}H_{16}N_5O_{13}P_3$
12. 5'-GTP $C_{10}H_{16}N_5O_{14}P_3$

Stainless steel	1.93 m	1 mm
Glass beads	50 μm	71°C
Gradient: 0.04M to 1.5M aqueous ammonium formate of pH 4.35		Polystyrene resin with strong, basic ion exchange groups chemically bonded
51 Bar	1 cm/sec	
Ultra-violet	260 nm	

First example of pellicular type support. Resin made by polymerising reactants which had been deposited on the glass beads. Ion exchange groups added by further reactions.
For abbreviations see 7.02.

7.02 RIBONUCLEOSIDE-2' AND 3'-MONOPHOSPHORIC ACIDS

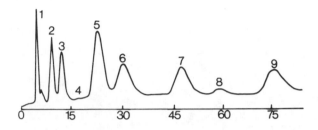

1. Bases
2. 2'-CMP $C_9H_{14}N_3O_8P$
3. 3'-CMP $C_9H_{14}N_3O_8P$
4. 2'-UMP $C_9H_{13}N_2O_9P$
5. 3'-UMP $C_9H_{13}N_2O_9P$

6. 2'-AMP $C_{10}H_{14}N_5O_7P$
7. 3'-AMP $C_{10}H_{14}N_5O_7P$
8. 2'-GMP $C_{10}H_{14}N_5O_8P$
9. 3'-GMP $C_{10}H_{14}N_5O_8P$

Stainless steel	1.93 m	1 mm
Glass beads	44-53 μm	60°C
Gradient: 0.05M to 0.35M aqueous KH_2PO_4		Polystyrene resin with strong, basic ion exchange groups chemically bonded
51 Bar	1 cm/sec	
Ultra-violet	260 nm	

Support described in 7.01. Abbreviations: C=Cytidine; U=Uridine; A=Adenosine; G=Guanosine; MP=monophosphoric acid; DP=diphosphoric acid and TP=triphosphoric acid.

Horvath, C.G., Preiss, B.A., Lipsky, S.R., *Analytical Chemistry,* **39**, 1422 (1967).

7.03 NUCLEOTIDES

1. Adenosine-5'-monophosphoric acid
   $C_{10}H_{14}N_5O_7P$
2. 3',5'-Cyclic adenosine monophosphate
   $C_{10}H_{12}N_5O_6P$
3. Inosine-5'-monophosphoric acid
   $C_{10}H_{13}N_4O_8P$

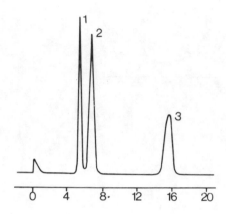

Stainless steel	3.0 m	1 mm
Glass beads	44-53 μm	75°C
0.013M hydrochloric acid		Polystyrene resin with strong, basic ion exchange groups chemically bonded
56 Bar	0.167 cm³/min	1 μl
Ultra-violet	254 nm	0.16 AUFS

Support was the same as used in 7.01.

Pennington, S.N., *Analytical Chemistry*, **43,** 1701 (1971). Copyright (1971) by the American Chemical Society. Reproduced by permission of the copyright owner.

7.04   RIBONUCLEOSIDE-2' AND 3'-MONOPHOSPHORIC ACIDS

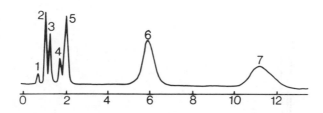

1. Bases
2. 2'-CMP $C_9H_{14}N_3O_8P$
3. 3'-CMP $C_9H_{14}N_3O_8P$
4. 2'-UMP $C_9H_{13}N_2O_9P$

5. 3'-UMP $C_9H_{13}N_2O_9P$
6. 3'-AMP $C_{10}H_{14}N_5O_7P$
7. 3'-GMP $C_{10}H_{14}N_5O_8P$

Stainless steel	1.0 m	2.1 mm
Zipax[R] SAX	20-37 μm	60°C
Water containing 0.006M phosphoric acid and 0.002M KH_2PO_4 of pH 3.75.		1% quaternary ammonium substituted polystyrene polymer
63 Bar	1.91 cm^3/min	2 μl
Ultra-violet	254 nm	

Note increase in speed to that of separation 7.02.

Kirkland, J.J., *Journal of Chromatographic Science,* **8,** 72 (1970). Reproduced by permission of Preston Technical Abstracts Co., Illinois.

7.05 NUCLEIC ACID BASES

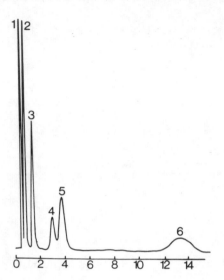

1. Uracil $C_4H_4N_2O_2$
2. Hypoxanthine $C_5H_4N_4O$
3. Guanine $C_5H_5N_5O$

4. Cytosine $C_4H_5N_3O$
5. Adenine $C_5H_5N_5$
6. 6-Methylaminopurine $C_6H_7N_5$

Stainless steel	1.0 m	2.1 mm
Zipax[R] SCX	20-37 μm	63°C
0.01N nitric acid		1% sulfonated fluorocarbon
51 Bar	2 cm^3/min	2 μl
Ultra-violet	254 nm	

Kirkland, J.J., *Journal of Chromatographic Science,* **8,** 72 (1970). Reproduced by permission of Preston Technical Abstracts Co., Illinois.

7.06 NUCLEOSIDE ANOMERS

1. 9-(4-*C*-cyclopropyl-β-D-*ribo*-tetrafuranosyl)adenine
   $C_{12}H_{15}N_5O_3$
2. 9-(4-*C*-cyclopropyl-α-D-*ribo*-tetrafuranosyl) adenine
   $C_{12}H_{15}N_5O_3$

Glass	0.50 m	2 mm
Zipax$^{(R)}$ SCX	20-37 μm	Ambient
0.025 sodium nitrate		1% sulfonated fluorocarbon
83 Bar	1.5 cm^3/min	20 μl
Ultra-violet	254 nm	0.08 AUFS

Chromatronix Inc., Liquid Chromatography Application No. 12. Reproduced by permission of Chromatronix Inc., Berkeley.

7.07 RIBONUCLEOSIDE MONOPHOSPHORIC ACIDS

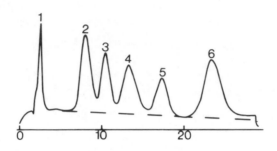

1. 5'-CMP $C_9H_{14}N_3O_8P$
2. 5'-UMP $C_9H_{13}N_2O_9P$
3. 5'-AMP $C_{10}H_{14}N_5O_7P$

4. 5'-IMP $C_{10}H_{13}N_4O_8P$
5. 3',5'-Cyclic AMP $C_{10}H_{12}N_5O_6P$
6. 5'-GMP $C_{10}H_{14}N_5O_8P$

Stainless steel	2.50 m	1 mm
Glass beads		70°C
Gradient: 0.01M KH_2PO_4 containing 0.01M H_3PO_4 (pH 2.6) to 0.15M KH_2PO_4 in 30 min.		Polystyrene resin with strong, basic ion exchange groups chemically bonded
56 Bar	0.4 cm³/min	10 μl
Ultra-violet	254 nm	0.08 AUFS

Abbreviations: I=Inosine; for others see 7.02.
Gradient produced by adding 1.22M KH_2PO_4, with complete mixing, to 50 ml of the starting mobile phase at half the rate that the mixture is being pumped through the column.

Shmukler, H.W., *Journal of Chromatographic Science*, **10**, 137 (1972). Reproduced by permission of Preston Technical Abstracts Co., Illinois.

7.08 FLAVIN NUCLEOTIDES

1. Riboflavin $C_{17}H_{20}N_4O_6$
2. Adenosine-5'-monophosphoric acid $C_{10}H_{14}N_5O_7P$
3. Flavin mononucleotide $C_{17}H_{21}N_4O_9P$
4. Flavin adenine dinucleotide $C_{27}H_{33}N_9O_{15}P_2$

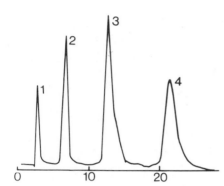

Stainless steel	2.50 m	1 mm
Glass beads		70°C
Gradient: 0.01M KH_2PO_4 containing 0.01M H_3PO_4 (pH 3.45) to 0.66M KCl containing 0.075M KH_2PO_4 in 3 min,		Polystyrene resin with strong, basic ion exchange groups chemically bonded
56 Bar	0.4 cm³/min	10 μl
Ultra-violet	254 nm	0.32 AUFS

Gradient produced by adding 2.20M KCl containing 0.25M KH_2PO_4, with complete mixing, to 20 ml of the starting mobile phase at half the rate that the mixture is being pumped through the column.

Shmukler, H.W., Journal of Chromatographic Science, **10**, 137 (1972). Reproduced by permission of Preston Technical Abstracts Co., Illinois.

8.01 RIFAMPIN AND RELATED COMPOUNDS

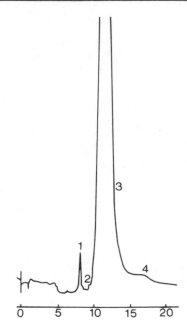

1. 3-Formyl Rifamycin SV $C_{38}H_{47}NO_{13}$
2. Desacetyl Rifampin $C_{41}H_{56}N_4O_{11}$
3. Rifampin $C_{43}H_{58}N_4O_{12}$
4. Rifampin quinone $C_{43}H_{56}N_4O_{12}$

Stainless steel	1.0 m	2.1 mm
Permaphase(R) ODS	37-44 μm	50°C
Gradient: water to methanol at 8% per minute		1% octadecylsilyl chemically bonded
70 Bar	1 cm³/min	
Ultra-violet	254 nm	

Schmit, J.A., Henry, R.A., Williams, R.C., Dieckman, J.F., *Journal of Chromatographic Science,* **9**, 645 (1971). Reproduced by permission of Preston Technical Abstracts Co., Illinois.

8.02 SULFONYLUREA ANTIDIABETICS

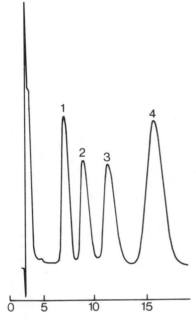

1. Chlorpropamide $C_{10}H_{13}ClN_2O_3S$
2. Tolazamide $C_{14}H_{21}N_3O_3S$
3. Tolbutamide $C_{12}H_{18}N_2O_3S$
4. Acetohexamide $C_{15}H_{20}N_2O_4S$

Stainless steel	1.0 m	2.1 mm
Zipax$^{(R)}$ HCP		Ambient
10% methanol in 0.01M aqueous monobasic sodium citrate	1% ethylene propylene copolymer	
35 Bar	0.36 cm^3/min	1 μl
Ultra-violet	254 nm	

8.03 DIPHENYLHYDANTOIN AND METABOLITE

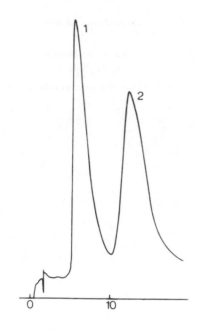

1. Diphenylhydantoin $C_{15}H_{12}N_2O_2$
2. 5-p-Hydroxyphenyl-5-phenylhydantoin $C_{15}H_{12}N_2O_3$

Stainless steel	3.0 m	1 mm
Glass beads		80°C
0.02M KH$_2$PO$_4$ of pH 4.5		Polystyrene resin with strong, basic ion exchange groups chemically bonded
100 Bar	0.4 cm^3/min	
Ultra-violet	254 nm	0.04 AUFS

At pH 3.5 peak shape was better. Sample was synthetic mixture.

Anders, M.W., Latorre, J.P., *Analytical Chemistry,* **42**, 1430 (1970). Copyright (1970) by the American Chemical Society. Reproduced by permission of the copyright owner.

8.04 4,4'-DIAMINODIPHENYL SULFONE AND METABOLITES

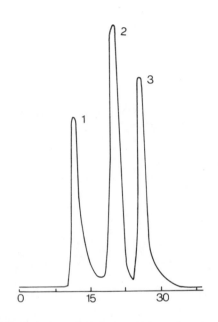

1. 4,4'-Diaminodiphenyl sulfone $C_{12}H_{12}N_2O_2S$
2. 4-Acetamido-4'-aminodiphenyl sulfone $C_{14}H_{14}N_2O_3S$
3. 4,4'-Diacetamidodiphenyl sulfone $C_{16}H_{16}N_2O_4S$

Glass	0.30 m	2.8 mm
Silica gel		Ambient
Ethyl acetate		None
7 Bar	0.25 cm^3/min	10 μl
Ultra-violet	280 nm	

Adsorbent was TLC Silica. Sample a urine extract.

Gordon, G.R., Peters, J.H., *Journal of Chromatography,* **47,** 269 (1970). Reproduced by permission of Elsevier, Amsterdam.

8.05 IMPURITIES IN 3,4',5-TRIBROMOSALICYLANILIDE

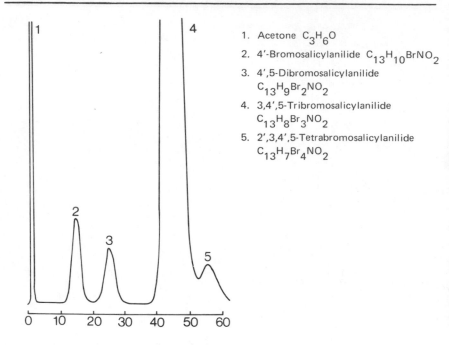

1. Acetone C_3H_6O
2. 4'-Bromosalicylanilide $C_{13}H_{10}BrNO_2$
3. 4',5-Dibromosalicylanilide $C_{13}H_9Br_2NO_2$
4. 3,4',5-Tribromosalicylanilide $C_{13}H_8Br_3NO_2$
5. 2',3,4',5-Tetrabromosalicylanilide $C_{13}H_7Br_4NO_2$

	0.50 m	2.8 mm
Bio-Rad[R] AGl-X2	37-74 μm	Ambient
Gradient: methanol to acetic acid		
	2.4 cm^3/min	10 μl
Ultra-violet	280 nm	0.2 AUFS

Gradient system was 'home made'. Time to complete gradient was about 60 minutes.
Acetone was solvent for sample.

Skelly, N.E., Crummett, W.B., *Journal of Chromatography,* **55**, 309 (1971). Reproduced by permission of Elsevier, Amsterdam.

8.06 BARBITURATES AND METABOLITES

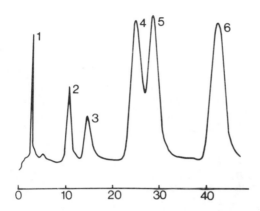

1. Ketohexobarbital $C_{12}H_{14}N_2O_4$
2. Hydroxyamobarbital $C_{11}H_{18}N_2O_4$
3. Impurity

4. Amobarbital $C_{11}H_{18}N_2O_3$
5. Phenobarbital $C_{12}H_{12}N_2O_3$
6. Hydroxyphenobarbital $C_{12}H_{12}N_2O_4$

Stainless steel	3.0 m	1 mm
Glass beads		80°C
Gradient: 0.0001M to 0.00028M sodium chloride in 50 min.		Polystyrene resin with strong, basic ion exchange groups chemically bonded
49-63 Bar	0.43 cm^3/min	
Ultra-violet	254 nm	0.08 AUFS

Gradient produced by adding 0.001M NaCl, with complete mixing to 50 ml of the starting mobile phase at half the rate that the mixture is being pumped through the column.
Sample was synthetic mixture.

8.07 VITAMIN STANDARDS

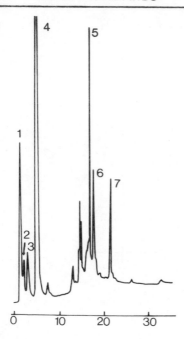

1. Riboflavin $C_{17}H_{20}N_4O_6$
2. Vitamin B_{12} $C_{63}H_{88}CoN_{14}O_{14}P$
3. Vitamin K_3 $C_{11}H_8O_2$
4. Vitamin C $C_6H_8O_6$
5. Vitamin D_2 $C_{28}H_{44}O$
6. Vitamin E $C_{29}H_{50}O_2$
7. Vitamin A $C_{20}H_{30}O$

Stainless steel	1.0 m	2.1 mm
Permaphase$^{(R)}$ ODS	20-37 μm	50°C
Gradient: water to methanol at 3% per minute		1% octadecylsilyl chemically bonded
53 Bar	0.9 cm^3/min	
Ultra-violet	254 nm	

Schmit, J.A., Henry, R.A., Williams, R.C., Dieckamn, J.F., *Journal of Chromatographic Science,* **9**, 645 (1971). Reproduced by permission of Preston Technical Abstracts Co., Illinois.

8.08 CLOPIDOL IN POULTRY FEED

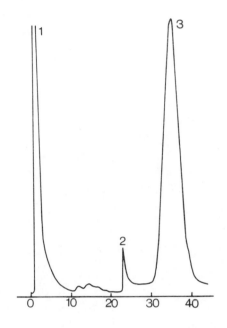

1. Feed extractable
2. Feed extractable
3. Clopidol $C_7H_7Cl_2NO$

	0.50 m	2.8 mm
Bio-Rad[(R)] AG1-X2	37-74 μm	Ambient
Gradient: methanol to 0.1% acetic acid in methanol		
	1.4 cm³/min	2 ml
Ultra-violet	268 nm	0.25 AUFS

2 ml of feed extract cleaned up by washing through a column of Al_2O_3
connected in series with the analytical column.
Eluted with 27 ml methanol after which Al_2O_3 column disconnected, gradient
started, and the sample chromatographed.
Resin was in acetate form.

Skelly, N.E., Cornier, R.F., *Journal of the AOAC,* **54,** 551 (1971). Reproduced by
permission of the Association of Official Analytical Chemists.

8.09 DIAZEPAM AND METABOLITIES

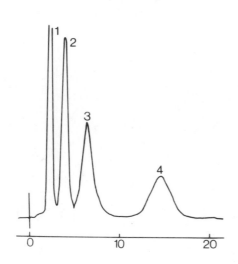

1. Diazepam $C_{16}H_{13}ClN_2O$
2. *N*-Demethylated metabolite $C_{15}H_{11}ClN_2O$
3. Hydroxylated metabolite $C_{16}H_{13}ClN_2O_2$
4. Oxazepam $C_{15}H_{11}ClN_2O_2$

Stainless steel	1.0 m	1 mm
Durapak$^{(R)}$ OPN	36-75 μm	Ambient
20% isopropanol in hexane		BOP chemically bonded
	1.0 cm^3/min	
Ultra-violet	254 nm	

Oxazepam is the product formed by the hydroxylation and demethylation of diazepam.

Scott, C.G., Bommer, P., *Journal of Chromatographic Science,* **8,** 446 (1970).
Reproduced by permission of Preston Technical Abstracts Co., Illinois.

8.10 COMPONENTS OF ANALGESIC TABLETS

1. Caffeine $C_8H_{10}N_4O_2$
2. Aspirin $C_9H_8O_4$
3. Paracetamol $C_8H_9NO_2$
4. Phenacetin $C_{10}H_{13}NO_2$
5. Salicylamide $C_7H_7NO_2$

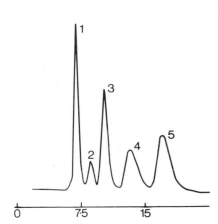

Stainless steel	3 m	1 mm
Glass beads	40 μm	60°C
Water containing 1.0M tris (hydroxmethyl) aminoethane adjusted to pH 9 with HC l		Polystyrene with strong, basic ion exchange groups chemically bonded
70 Bar	0.61 cm/sec	
Ultra-violet	254 nm	

Chromatograms of 12 commercial tablets also given

Stevenson, R.L., Burtis, C.A., *Journal of Chromatography*, **69**, 253 (1971). Reproduced by permission of Elsevier, Amsterdam.

8.11 MULTICOMPONENT ANALGESIC TABLET

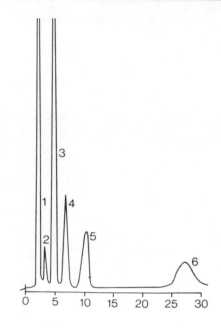

1. Codeine phosphate $C_{18}H_{24}NO_4P$
2. Caffeine $C_8H_{10}N_4O_2$
3. Phenacetin $C_{10}H_{13}NO_2$
4. Aspirin $C_9H_8O_4$
5. Benzoic acid $C_7H_6O_2$
6. Phenobarbital $C_{12}H_{12}N_2O_3$

Stainless steel	1.0 m	2.1 mm
Zipax(R) SAX	25-37 μm	Ambient
Water containing 0.005M sodium nitrate of pH 9		1% quaternary ammonium substituted polystyrene polymer
84 Bar	1.2 cm^3/min	
Ultra-violet	254 nm	

Benzoic acid used as internal standard

Schmit, J.A., 'Applications of High Speed Liquid Chromatography Using CSP Supports' in *Modern Practice of Liquid Chromatography* edited by Kirkland, J.J. Copyright (C), John Wiley and Sons, Inc., (1971). Reproduced by permission of John Wiley and Sons, Inc.

8.12 SULFONAMIDES

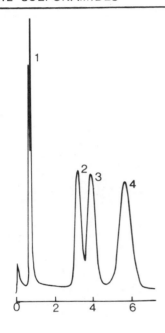

1. *N,N*-Dimethyl-*p*-toluenesulfonamide
   $C_9H_{13}NO_2S$
2. *N,N*-Dimethylbenzenesulfonamide
   $C_8H_{11}NO_2S$
3. *N*-Methyl-*p*-toluenesulfonamide
   $C_8H_{11}NO_2S$
4. *N*-Phenyl-p-toluenesulfonamide
   $C_{13}H_{13}NO_2S$

Stainless steel	1.0 m	2.1 mm
Permaphase(R) ETH	20-37 μm	27°C
5% chloroform in hexane		0.77% ether chemically bonded
60 Bar	2.66 cm^3/min	
Ultra-violet	254 nm	

Kirkland, J.J., De Stefano, J.J., *Journal of Chromatographic Science,* **8**, 309 (1970).
Reproduced by permission of Preston Technical Abstracts Co. Illinois.

8.13 VITAMIN STANDARDS

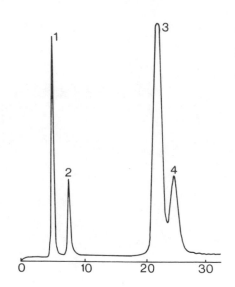

1. Vitamin A palmitate $C_{36}H_{60}O_2$
2. Vitamin E $C_{29}H_{50}O_2$
3. Vitamin D_2 $C_{28}H_{44}O$
4. Vitamin A $C_{20}H_{30}O$

	2.4 m	2.3 mm
Corasil$^{(R)}$ II	37-50 μm	Ambient
25% iso-octane in chloroform		None
	0.8 cm^3/min	
Ultra-violet	254 nm	0.16 AUFS

Reproduced by courtesy of Waters Associates, Inc.

8.14 PHENETHYLAMINES

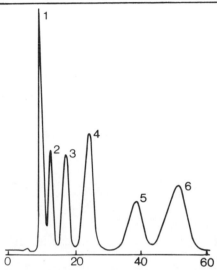

1. Ephedrine $C_{10}H_{15}NO$

2. Dexedrine $C_9H_{15}NO_4S$

3. N-Methyl-3,4-(methylenedioxy) phenethylamine $C_{10}H_{13}NO_2$

4. STP $C_{12}H_{19}NO_2$

5. Benzphetamine $C_{17}H_{21}N$

6. α-Benzyl-3,4-dimethylphenethylamine $C_{17}H_{21}N$

Glass	0.50 m	3 mm
Durrum(R) DA-X4		55°C
Water containing 0.2M sodium nitrate adjusted to pH 3.15		
35 Bar	0.26 cm³/min	30 µl
Ultra-violet	254 nm	0.16 AUFS

Chromatronix Inc., Liquid Chromatography Application Number 11. Reproduced by permission of Chromatronix Inc., Berkeley.

8.15 ASTHMA AND HAY FEVER TABLET

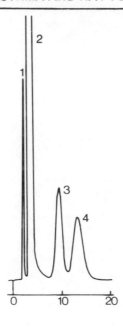

1. Ephedrine $C_{10}H_{15}NO$
2. Theophyline $C_7H_8N_4O_2$
3. Benzoic acid $C_7H_6O_2$
4. Phenobarbital $C_{12}H_{12}N_2O_3$

Stainless steel	1.50 m	2 mm
Zipax$^{(R)}$ SAX	29-37 μm	37°C
Water containing 0.01M sodium nitrate adjusted to pH 5.7		1% quaternary ammonium substituted polystyrene polymer
124 Bar	1 cm^3/min	20 μl
Ultra-violet	254 nm	0.08 AUFS

Tablet dissolved in 50 ml methanol and injected with no further treatment. Benzoic acid used as internal standard.

Chromatronix Inc., Liquid Chromatography Application Number.13. Reproduced by permission of Chromatronix Inc., Berkeley.

8.16 VITAMINS

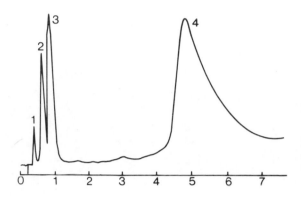

1. β-Carotene $C_{40}H_{56}$
2. Vitamin A acetate $C_{22}H_{32}O_2$
3. Vitamin E acetate $C_{31}H_{52}O_2$
4. Vitamin D_3 $C_{27}H_{44}O$

	0.10 m	3 mm
Silica gel		20°C
Iso-octane: hexane: ethyl-*n*-butylether: ether in the ratio 30:10:2:1	None	
	1.6 cm³/min	
Ultra-violet	270 nm	

Siemens' liquid chromatography bulletin S 200. Reproduced by permission of Siemens, AG, Karlsruhe.

8.17 P-HYDROXYACETANILIDE METABOLITES

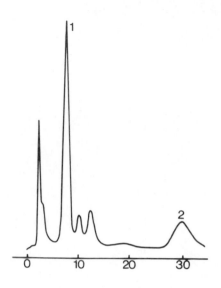

1. Acetanilide glucuronide
   $C_{16}H_{17}NO_8$
2. Acetanilide sulfate $C_8H_9NO_5S$

Stainless steel	2.5 m	1 mm
Glass beads		$80^{\circ}C$
Gradient: 0.001M formic acid to 0.001M formic acid containing 0.6M KC1 in 40 min.		Polystyrene with strong, basic ion exchange groups chemically bonded
56-71 Bar	0.5 cm^3/min	1 μl
Ultra-violet	254 nm	0.32 AUFS

Direct injection of urine sample of adult male after treatment with p-hydroxyacetanilide.

Anders, M.W., Latorre, J.P., *Journal of Chromatography,* **55,** 409 (1971). Reproduced by permission of Elsevier, Amsterdam.

8.18 IMPURITIES IN METHYL ESTER OF BENZYLPENICILLIN

1. Methyl ester of benzylpenicillin
    $C_{17}H_{20}N_2O_4S$

Stainless steel	1.0 m	2.1 mm
Zipax$^{(R)}$ PAM	20-37 μm	Ambient
5% hexane in ethanol		1% polyamide
63 Bar	0.9 cm^3/min	
Ultra-violet	254 nm	

Schmit, J.A., 'Applications of High Speed Liquid Chromatography Using CSP Supports' in *Modern Practice of Liquid Chromatography* edited by Kirkland, J.J. Copyright (C), John Wiley and Sons, Inc., (1971). Reproduced by permission of John Wiley and Sons, Inc.

8.19 VITAMIN A AND VITAMIN A ACETATE

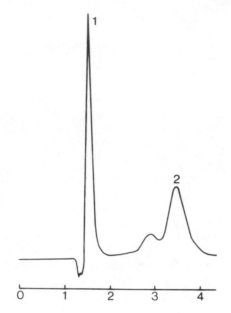

1. Vitamin A acetate $C_{22}H_{32}O_2$
2. Vitamin A $C_{20}H_{30}O$

	0.50 m	2.3 mm
Corasil$^{(R)}$ II	37-50 μm	Ambient
Chloroform		None
	0.75 cm^3/min	5 μl
Refractive index		

Bombaugh, K.J., Levangie, R.F., King, R.N., Abrahams, L., *Journal of Chromatographic Science,* **8**, 657 (1970). Reproduced by permission of Preston Technical Abstracts Co., Illinois.

8.20 METOLAZONE IN URINE

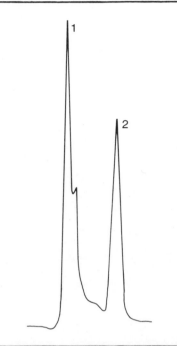

1. Unknown
2. Metolazone $C_{16}H_{16}ClN_3O_3S$

	1.80 m	3.2 mm
Durapak(R) PEG 400	36-75 μm	Ambient
5% isopropanol in chloroform		Polyethylene glycol chemically bonded
21 Bar	1 cm³/min	25 μl
Ultra-violet	254 nm	

Urine saturated with KH_2PO_4, extracted with chloroform, evaporated to dryness, and made up to 100 μl with tetrahydrofuran. A 5ml sample of urine was used.

Hinsvark, O.N., Zazulak, W., Cohen, A.I., *Journal of Chromatographic Science,* **10**, 379 (1972). Reproduced by permission of Preston Technical Abstracts Co., Illinois.

9.01   OESTROGENS

1. Oestrone $C_{18}H_{22}O_2$
2. Oestradiol $C_{18}H_{24}O_2$
3. Oestriol $C_{18}H_{24}O_3$

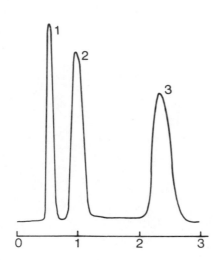

	0.088 m	3 mm
Chromosorb G	5-10 μm	Ambient
Iso-octane:ethanol:water in the molar ratio 80.5:17.7:1.9		Iso-octane:ethanol:water in the molar ratio 9.1:68.0:22.9
	1 cm^3/min	1 μl
Ultra-violet	270 nm	

Ecker, E., *Chemiker-Zeitung,* **95,** 511 (1971). Reproduced by permission of Chemiker-Zeitung, Heidelberg, Redaktion.

9.02 ADRENAL CORTICOSTEROIDS

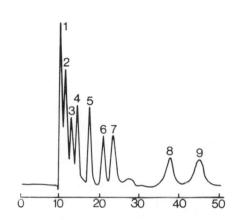

1. 6β-Hydroxycortisone $C_{21}H_{28}O_6$
2. Aldosterone $C_{21}H_{28}O_5$
3. Cortisone $C_{21}H_{28}O_5$
4. 11-Dehydrocorticosterone $C_{21}H_{28}O_4$
5. Corticosterone $C_{21}H_{30}O_4$
6. 11-Deoxycortisol $C_{21}H_{30}O_4$
7. Cortisone 21-acetate $C_{23}H_{30}O_6$
8. 4-Pregene-20β,21-diol-3-one
   $C_{21}H_{32}O_3$
9. Deoxycorticosterone $C_{21}H_{30}O_3$

Glass	0.485 m	2 mm
Trifluoroethylene	<44 μm	Ambient
Water		23% Amberlite[R] LA-1
	0.10-0.44 cm^3/min	1-10 μl
Ultra-violet	254 nm	

Amberlite[R] LA-1 is a weakly basic high molecular weight secondary amine.
The flow rate was changed in the following way: Start:0.10; at 10 min:0.13: at
16 min:0.21; at 20 min:0.26 and finally at 28 min to 0.44 om^3/min.

9.03 STEROIDS

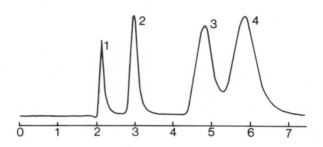

1. 6β-Hydroxycortisone $C_{21}H_{28}O_6$
2. Cortisol $C_{21}H_{30}O_5$

3. 4-Pregnene-11β,17α-diol-3,20-dione $C_{21}H_{30}O_4$
4. 4-Pregnene-17α-ol-3,11,20-trione $C_{21}H_{28}O_4$

Glass	0.485 m	2 mm
Trifluoroethylene	<44 μm	Ambient
Water		23% Amberlite[R] LA-1
31 Bar	0.49 cm³/min	
Ultra-violet	245 nm	

Amberlite[R] LA-1 is a weakly basic high molecular weight secondary amine.

Siggia, S., Dishman, R.A., *Analytical Chemistry,* **42,** 1223 (1970). Copyright (1970) by the American Chemical Society. Reproduced by permission of the copyright owner.

9.04 INSECT-MOLTING HORMONES

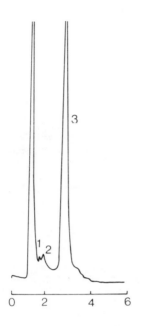

1. Ponasterone A $C_{27}H_{44}O_6$
2. Ponasterone B $C_{27}H_{44}O_6$
3. β-Ecdysone $C_{27}H_{44}O_6$

Stainless steel	1.0 m	2.1 mm
Zipax$^{(R)}$	37-44 μm	Ambient
10% Tetrahydrofuran in heptane		1% BOP
42 Bar	1 cm^3/min	
Ultra-violet	254 nm	

Schmit, J.A., 'Applications of High Speed Liquid Chromatography Using CSP Supports' in *Modern Practice of Liquid Chromatography* edited by Kirkland, J.J. Copyright (C), John Wiley and Sons, Inc. (1971). Reproduced by permission of John Wiley and Sons, Inc.

9.05 STEROIDS AND DERIVATIVES

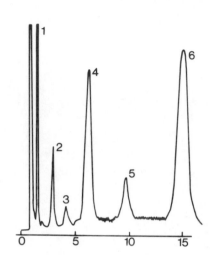

1. Progesterone $C_{21}H_{30}O_2$
2. 4-Androstene-3,17-dione $C_{19}H_{26}O_2$
3. Testosterone $C_{19}H_{28}O_2$
4. Dehydro*epi*androsterone, dinitro-phenylhydrazine derivative $C_{25}H_{32}N_4O_5$
5. 5α-Androstane-3β-ol-17-one, dinitro-phenylhydrazine derivative $C_{25}H_{34}N_4O_5$
6. Δ^1-Testosterone $C_{19}H_{26}O_2$

Stainless steel	1.0 m	2.1 mm
Zipax$^{(R)}$	20-37 μm	Ambient
Hexane		1.5% Polyethylene glycol 400
34.5 Bar	0.9 cm/sec	10 μl
Ultra-violet	254 nm	0.04 & 0.01 AUFS

Derivatisation of compounds 4 and 5 is necessary so that they have similar extinction coefficients to the other steroids.

Done, J.N., Knox, J.H., unpublished results

9.06 ANDROGENS

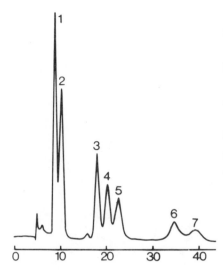

1. 4-Androstene-3,11,17-trione
   $C_{19}H_{24}O_3$
2. 4-Androstene-11β-ol-3,17-dione
   $C_{19}H_{26}O_3$
3. Δ1,4-Androstadiene-17β-ol-3-one
   $C_{19}H_{26}O_2$
4. 19-Nor-4-androstene-3,17-dione
   $C_{18}H_{24}O_2$
5. 19-Nor-testosterone $C_{18}H_{26}O_2$
6. 4-Androstene-3,17-dione $C_{19}H_{26}O_2$
7. Testosterone $C_{19}H_{28}O_2$

Glass	0.485 m	2 mm
Trifluoroethylene	44 μm	Ambient
Water		23% Amberlite[R] LA-1
	0.17 & 0.49 cm^3/min	
Ultra-violet	245 nm	

Amberlite[R] LA-1 is a weakly basic high molecular weight secondary amine, The flow rate was changed after 11 minutes.

9.07 STEROIDS

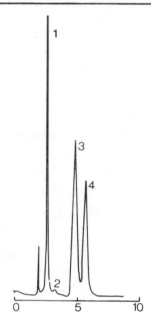

1. Progesterone $C_{21}H_{30}O_2$
2. Androsterone $C_{19}H_{30}O_2$
3. Testosterone $C_{19}H_{28}O_2$
4. 19-Nor-testosterone $C_{18}H_{26}O_2$

Stainless steel	1.0 m	2.1 mm
Zipax$^{(R)}$	37-44 μm	Ambient
Heptane		1% BOP
42 Bar	1 cm^3/min	
Ultra-violet	254 nm	0.32 AUFS

The UV detector is insensitive to androsterone. Detection limit about 1 μg. For other components it is about 10 ng. Separation using refractive index detection is also shown.

Henry, R.A., Schmit, J.A., Dieckman, J.F., *Journal of Chromatographic Science,* **9,** 513 (1971). Reproduced by permission of Preston Technical Abstracts Co., Illinois.

9.08 IMPURITIES IN PREDNISOLONE

1. Prednisolone $C_{21}H_{28}O_5$

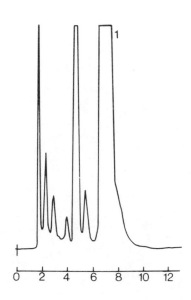

Stainless steel	1.0 m	2.1 mm
Zipax$^{(R)}$	37-44 μm	Ambient
7.5% ethylene chloride in heptane		1% ethylene glycol
75 Bar	1 cm^3/min	2 μl
Ultra-violet	254 nm	

Dupont 820 Liquid Chromatograph Bulletin. Reproduced by permission of Du Pont Instrument Products Division, Wilmington.

9.09 CORTICOSTEROIDS

1. Melengestrol acetate $C_{25}H_{32}O_4$
2. 6α-Methylprednisolone acetate $C_{24}H_{32}O_6$
3. 6α-Methylprednisolone $C_{22}H_{30}O_5$
4. 9α-Fluoroprednisolone acetate $C_{23}H_{29}FO_6$

Stainless steel	1.0 m	2.1 mm
Zipax (R)	37-44 μm	Ambient
20% tetrahydrofuran in heptane		1% BOP
42 Bar	1 cm^3/min	
Ultra-violet	254 nm	0.08 AUFS

Henry, R.A., Schmit, J.A., Dieckman, J.F., *Journal of Chromatographic Science,* **9**, 513 (1971). Reproduced by permission of Preston Technical Abstracts Co., Illinois.

9.10 CORTICOSTEROIDS

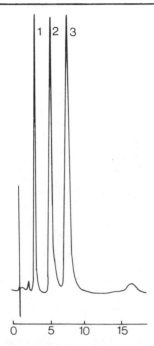

1. Prednisolone acetate $C_{23}H_{30}O_6$
2. Prednisone $C_{21}H_{26}O_5$
3. Prednisolone $C_{21}H_{28}O_5$

Stainless steel	1.0 m	2.1 mm
Zipax$^{(R)}$	37-44 μm	Ambient
20% tetrahydrofuran in heptane		1% BOP
42 Bar	1 cm^3/min	
Ultra-violet	254 nm	0.08 AUFS

Henry, R.A., Schmit, J.A., Dieckman, J.F., *Journal of Chromatographic Science,* **9,** 513 (1971) Reproduced by permission of Preston Technical Abstracts Co., Illinois.

9.11 SYNTHETIC OESTROGENS

1. Diethylstilbestrol dipropionate $C_{24}H_{28}O_4$
2. Dienestrol diacetate $C_{22}H_{22}O_4$
3. Diethylstilbestrol $C_{18}H_{20}O_2$
4. Dienestrol $C_{18}H_{18}O_2$

	0.50 m	3 mm
Sil-X$^{(R)}$	36-45 μm	Ambient
7% tetrahydrofuran in hexane		None
	1 cm^3/min	
Ultra-violet		

9.12 OESTROGENS

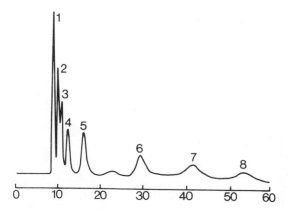

1. Oestradiol-17α-glucosidur-
   onic acid $C_{24}H_{32}O_8$
2. Oestriol $C_{18}H_{24}O_3$
3. 16-Keto-oestrone $C_{18}H_{20}O_3$
4. 16-Keto-oestradiol $C_{18}H_{22}O_3$

5. 16-*epi*Oestriol $C_{18}H_{24}O_3$
6. Equilinen $C_{18}H_{18}O_2$
7. Oestradiol $C_{18}H_{24}O_2$
8. Oestrone $C_{18}H_{22}O_2$

Glass	0.485 m	2 mm
Trifluoroethylene	44 μm	Ambient
Water adjusted to pH 11.5 with sodium hydroxide		28% Amberlite[R] LA-1
	0.12-0.49 cm³/min	
Ultra-violet	280 nm	

Amberlite[R] LA-1 is a weakly basic high molecular weight secondary amine.
The flow was changed in the following way: Start:0.12; at 10 min:0.145;
at 14 min:0.19 and finally at 20 min to 0.10 cm³/min.

Siggia, S., Dishman, R.A., *Analytical Chemistry*, **42**, 1223 (1970). Copyright (1970) by
the American Chemical Society. Reproduced by permission of the copyright owner.

9.13 OESTRADIOL IN SESAME OIL

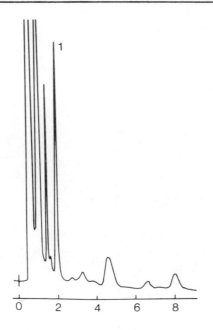

1. Oestradiol $C_{18}H_{24}O_2$

Stainless steel	1.0 m	2.1 mm
Zipax$^{(R)}$	37-44 μm	Ambient
5% tetrahydrofuran in heptane		1% BOP
42 Bar	1 cm^3/min	
Ultra-violet	254 nm	

9.14 CARDIAC GLYCOSIDES

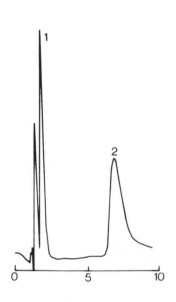

1. Digoxin $C_{41}H_{64}O_{14}$
2. Digitoxin $C_{41}H_{64}O_{13}$

Stainless steel	1.0 m	2.1 mm
Zipax$^{(R)}$ ANH	37–44 μm	40°C
2.5% methanol in water		1% cyanoethylsilicone polymer
42 Bar	0.5 cm^3/min	
Ultra-violet	254 nm	0.16 AUFS

Henry, R.A., Schmit, J.A., Dieckman, J.F., *Journal of Chromatographic Science,* **9**, 513 (1971). Reproduced by permission of Preston Technical Abstracts Co., Illinois.

10.01 CAROTENOIDS IN SPINACH EXTRACT

1. Carotenes $C_{40}H_{56}$
2. Lutein $C_{40}H_{56}O_2$
3. Violaxanthin $C_{40}H_{56}O_4$
4. Neoxanthin $C_{40}H_{56}O_4$

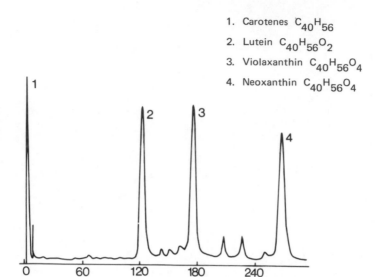

Stainless steel	0.135 m	6.35 mm
Zinc carbonate		15°C
Gradient: hexane to a maximum of 23.5% t-amyl alcohol in hexane		None
	0.5 cm³/min	50 μl
Visible	440 nm	1.0 AUFS

Gradient system consisted of 3 vessels containing hexane, 1.1% t-amyl alcohol in hexane and 23.5% t-amyl alcohol in hexane, respectively. Further details not given. An antioxidant was added to the mobile phase to prevent decomposition of the carotenoids.

Stewart, I., Wheaton, T.A., *Journal of Chromatography,* **55,** 325 (1971). Reproduced by permission of Elsevier, Amsterdam.

10.02   CAROTENES IN CARROT EXTRACT

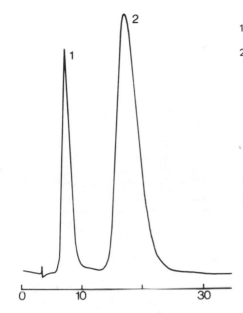

1. α-Carotene $C_{40}H_{56}$

2. β-Carotene $C_{40}H_{56}$

Glass	0.280 m	3.2 mm
Magnesium oxide		Ambient
5% t-amyl alcohol in hexane		None
	1.0 cm^3/min	50 μl
Visible	440 nm	0.8 AUFS

Stewart, I., Wheaton, T.A., *Journal of Chromatography,* **55**, 325 (1971). Reproduced by permission of Elsevier, Amsterdam.

10.03   ULTRA-VIOLET-ABSORBING CONSTITUENTS OF HUMAN URINE

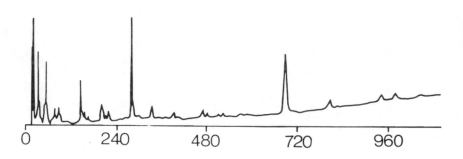

Stainless steel	1.00 m	2.4 mm
Bio-Rad(R) Aminex BRX	12-15 μm	21°C and 60°C
Gradient: 0.015M to 6.0M sodium acetate of pH 4.4.		
70-112 Bar	0.13 cm³/min	200 μl
Ultra-violet	254 nm	1.28 AUFS

Temperature changed after 240 minutes. Start of gradient delayed for 30 minutes. Time to complete was 24 hours.

Burtis, C.A., *Journal of Chromatography,* **52,** 97 (1970). Reproduced by permission of Elsevier, Amsterdam.

10.04   AFLATOXINS

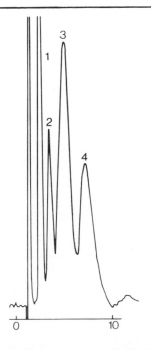

1. Aflatoxin B_1 $C_{17}H_{12}O_6$
2. Aflatoxin B_2 $C_{17}H_{14}O_6$
3. Aflatoxin G_1 $C_{17}H_{12}O_7$
4. Aflatoxin G_2 $C_{17}H_{14}O_7$

Glass	0.50 m	2 mm
Corasil$^{(R)}$ II	37-50 μm	Ambient
20% iso-octane in chloroform		None
11 Bar	0.7 cm^3/min	40 μl
Ultra-violet	254 nm	0.02 AUFS

Chromatronix Inc., Liquid Chromatography Applications Number 8. Reproduced by permission of Chromatronix Inc., Berkeley.

10.05   STRYCHNOS ALKALOIDS

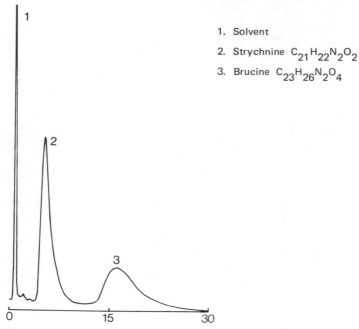

1. Solvent
2. Strychnine $C_{21}H_{22}N_2O_2$
3. Brucine $C_{23}H_{26}N_2O_4$

Glass	0.50 m	1 mm
Corasil$^{(R)}$ I	37-50 μm	Ambient
9% ethanol in heptane		1.1% Poly G-300
21 Bar	0.27 cm^3/min	2 μl
Ultra-violet	270 nm	0.08 AUFS

10.06   WORLD STANDARD PYRETHRIN EXTRACT

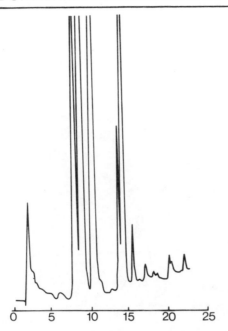

Stainless steel	1.00 m	2.1 mm
Permaphase(R) ODS	37-44 μm	50°C
Gradient: 30% methanol in water to methanol at 3% per minute		1% octadecylsilyl chemically bonded
70 Bar	1.5 cm³/min	
Ultra-violet	254 nm	0.32 AUFS

Schmit, J.A., Henry, R.A., Williams, R.C., Dieckman, J.F., *Journal of Chromatographic Science,* **9**, 645 (1971). Reproduced by permission of Preston Technical Abstracts Co., Illinois.

10.07   COUMARIN AND DERIVATIVES

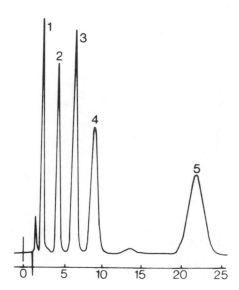

1. 7-Hydroxy-6-methoxycoumarin
   $C_{10}H_8O_4$
2. Coumarin $C_9H_6O_2$
3. 3,4-Dihydrocoumarin $C_9H_8O_2$
4. 6-Methylcoumarin $C_{10}H_8O_2$
5. 7-Ethoxy-4-methylcoumarin
   $C_{12}H_{12}O_3$

Stainless steel	1.00 m	2.1 mm
Zipax$^{(R)}$ ANH	37-44 μm	40°C
Water		1% cyanoethylsilicone polymer
84 Bar	1.5 cm^3/min	
Ultra-violet	254 nm	

10.08   LOW MOLECULAR WEIGHT FRACTION OF HASHISH EXTRACT

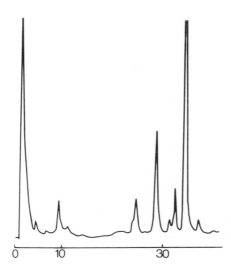

Stainless steel	1.00 m	2.1 mm
Permaphase(R) ODS	37-44 μm	50°C
Gradient: water to methanol at 2% per minute.		1% octadecylysilyl chemically bonded
	1 cm³/min	
Ultra-violet	254 nm	

Gel Permeation Chromatography used to obtain the fraction of the crude extract that corresponded to the molecular weight of tetrahydrocannabinol (a component of Cannabis resin).

Schmit, J.A., Henry, R.A., Williams, R.C., Dieckman, J.F., *Journal of Chromatographic Science,* **9,** 645 (1971). Reproduced by permission of Preston Technical Abstracts Co., Illinois.

10.09   PRINCIPAL ALKALOIDS OF AN IPECAC EXTRACT

1. Emetine $C_{28}H_{38}N_2O_4$
2. Cephaeline $C_{29}H_{40}N_2O_4$

Stainless steel	1.00 m	2.1 mm
Zipax$^{(R)}$ ANH	37-44 μm	40°C
Heptane		1% cyanoethylsilicone polymer
63 Bar	2 cm^3/min	
Ultra-violet	254 nm	

10.10   PURINE ALKALOIDS

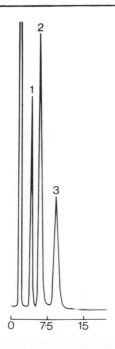

1. Caffeine $C_8H_{10}N_4O_2$
2. Theophylline $C_7H_8N_4O_2$
3. Theobromine $C_7H_8N_4O_2$

Glass	1.00 m	1 mm
Corasil$^{(R)}$ II	37-50 μm	Ambient
9% ethanol in heptane		1.1% Poly G-300

21 Bar	0.27 cm^3/min	2 μl
Ultra-violet	270 nm	0.08 AUFS

Synthetic mixture. Using Corasil$^{(R)}$ I, elution order was 1,3,2.

10.11   OPIUM ALKALOIDS

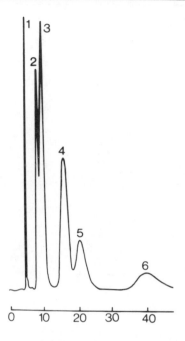

1. Morphine $C_{17}H_{19}NO_3$
2. Codeine $C_{18}H_{21}NO_3$
3. Papaverine $C_{20}H_{21}NO_4$
4. Thebaine $C_{19}H_{21}NO_3$
5. Cryptopine $C_{21}H_{23}NO_5$
6. Narcotine $C_{22}H_{23}NO_7$

Stainless steel	1.00 m	2.1 mm
Zipax$^{(R)}$ SCX	37-44 μm	Ambient
1% propanol and 4% acetonitrile in water containing 0.2M KNO_3 and 0.2M NaOH adjusted to pH 9.45 with phosphoric acid		1% sulfonated fluorocarbon
28 Bar	0.37 cm/sec	5 μl
Ultra-violet	254 nm	0.5 AUFS

Jurand, J., Knox, J.H., unpublished results

11.01   GALLIUM, INDIUM AND THALLIUM

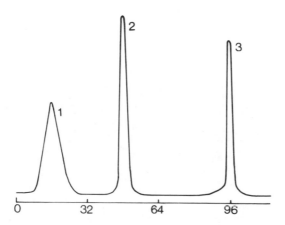

1. Gallium Ga   3. Thallium Tl

2. Indium In

Glass	0.12 m	10 mm
Amberlyst$^{(R)}$ XAD-2	149-177 μm	Ambient
Gradient: stepwise; gallium eluted with 5M HBr; indium with 1M HBr and thallium with 3M HNO$_3$ and isopropyl ether		Isopropyl ether
	1.0–1.5 cm^3/min	
Titration		

Although not strictly a high speed separation because the detection involved taking fractions and analysing by titration, this system would be easy to adapt so that the polarograph detector (see 0.04) could be utilised.

Fritz, J.S., Frazee, R.T., Latwesen, G.L., *Talanta,* **17,** 857 (1970). Reproduced by permission of Microforms International Marketing Corporation, New York.

11.02   IMPURE SODIUM NITRITE

1. Sodium nitrite $NaNO_2$
2. Sodium nitrate $NaNO_3$

Stainless steel	1.00 m	2.1 mm
Zipax$^{(R)}$ SAX	25-37 μm	Ambient
Water		1% quaternary ammonium substituted polystyrene polymer
84 Bar	1.5 cm^3/min	2 μl
Ultra-violet	254 nm	

Schmit, J.A., Henry, R.A., *Chromatographia*, **3**, 497 (1970). Reproduced by permission of R. Henry.

11.03   2,3-DIMETHYLNAPHTHALENE METAL CARBONYLS

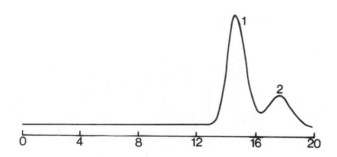

1.   Tricarbonyl(η-1,2,3,4,4a,8a-2,3-dimethylnaphthalene)chromium $C_{15}H_{12}CrO_3$

2.   Tricarbonyl(η-4a,5,6,7,8,8a-2,3-dimethylnaphthalene)chromium $C_{15}H_{12}CrO_3$

Glass	1.00 m	3.5 mm
Porasil$^{(R)}$ C	36-75 μm	26.5°C
Iso-octane		Carbowax$^{(R)}$ 400 chemically bonded
	1.85 cm^3/min	
Ultra-violet	350 nm	

Greenwood, J.M., Veening, H., Willeford, B.R., *Journal of Organometallic Chemistry*, **38**, 345 (1972). Reproduced by permission of Elsevier, Amsterdam.

11.04   ARENE TRICARBONYLCHROMIUM COMPLEXES

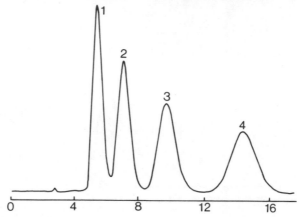

1. Mesitylenetricarbonylchromium
   $C_{12}H_{12}CrO_3$

2. m-Xylenetricarbonylchromium
   $C_{11}H_{10}CrO_3$

3. Toluenetricarbonylchromium
   $C_{10}H_8CrO_3$

4. Benzenetricarbonylchromium
   $C_9H_6CrO_3$

Glass	0.55 m	3.5 mm
Porasil[(R)].C	36-75 μm	30°C
Iso-octane		Carbowax[(R)] 400 chemically bonded
	1.85 cm^3/min	10 μl
Ultra-violet	320 nm	0.8 AUFS

Veening, H., Greenwood, J.M., Shanks, W.H., Willeford, B.R., *Chemical Communications*, 1305 (1969). Reproduced by permission of The Chemical Society, London.

11.05   SUBSTITUTED BENZIMIDAZOLES

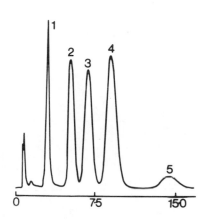

1. 2-Aminobenzimidazole $C_7H_7N_3$
2. 5,6-Dichlorobenzimidazole
   $C_7H_4Cl_2N_2$
3. 5,6-Dimethyl-2-benzimidazolemethanol
   $C_{10}H_{12}N_2O$
4. 5,6-Dimethylbenzimidazole $C_9H_{10}N_2$
5. 2-Amino-5,6-dimethylbenzimidazole
   $C_9H_{11}N_3$

Stainless steel	1.00 m	2.1 mm
Zipax$^{(R)}$ SCX	20-37 μm	60°C
Water containing 0.025N tetramethylammonium nitrate and HNO$_3$ of pH 1.74		1% sulfonated fluorocarbon
23 Bar	2.05 cm^3/min	10 μl
Ultra-violet	254 nm	0.05 AUFS

Kirkland, J.J., *Journal of Chromatographic Science,* **7**, 361 (1969). Reproduced by permission of Preston Technical Abstracts Co., Illinois.

11.06   SULPHUR HETEROCYLES

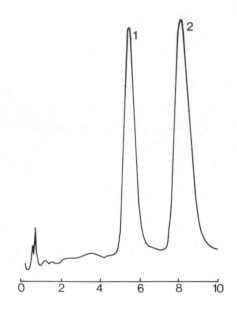

1. 4-Neopentyl-1,2-dithiole-3-thione $C_8H_{12}S_3$
2. 5-t-Butyl-4-methyl-1,2-dithiole-3-thione $C_8H_{12}S_3$

Glass	0.90 m	2.1 mm
Corasil$^{(R)}$ II	37-44 μm	Ambient
Hexane		None
41 Bar	2.7 cm/sec	1 μl
Ultra-violet	254 nm	0.08 AUFS

Done, J.N., Knox, J.H., unpublished results.

11.07   SYNTHETIC ORGANICS

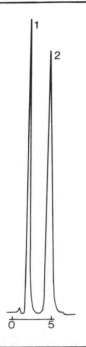

1. 1-(3,4-Dibenzyloxyphenyl)-2-
   nitro-*trans*-prop-1-ene $C_{23}H_{21}NO_4$
2. 3,4-Dibenzyloxybenzaldehyde
   $C_{21}H_{18}O_3$

Glass	0.50 m	2 mm
Corasil$^{(R)}$ II	37-50 μm	Ambient
10% chloroform in iso-octane		None
13 Bar	1 cm^3/min	40 μl
Ultra-violet	254 nm	0.64 AUFS

Chromatronix Inc., Liquid Chromatography Application Number 4. Reproduced by
permission of Chromatronix Inc., Berkeley.

11.08   SULPHUR HETEROCYCLES

1. 2,5-Diphenyl-6a-selenathiophthen
   $C_{17}H_{12}S_2Se$

2. 2,5-Diphenyl-6a-thiathiophthen
   $C_{17}H_{12}S_3$

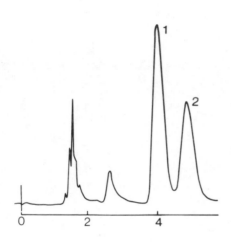

Stainless steel	1.00 m	2.1 mm
Corasil(R) II	37-44 μm	Ambient
Hexane		None
14 Bar	1.28 cm/sec	2 μl
Ultra-violet	254 nm	0.16 AUFS

Done, J.N., Knox, J.H., unpublished results

11.09   ORGANIC REACTION MIXTURE

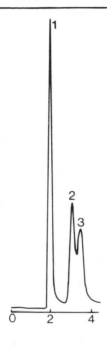

1. p-Chlorophenyl o-nitrophenyl sulphide $C_{12}H_8ClNO_2S$
2. 3-Chlorophenothiazine $C_{12}H_8ClNS$
3. 2-Chlorophenothiazine $C_{12}H_8ClNS$

Glass	1.00 m	2.1 mm
Zipax(R)	37-44 μm	Ambient
Cyclohexane		1% BOP
41.5 Bar	0.86 cm/sec	2 μl
Ultra-violet	254 nm	0.04 AUFS

Done, J.N., Knox, J.H., unpublished results

11.10   1,2-BENZODIAZEPINES

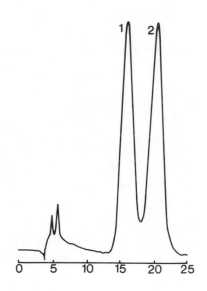

1. 1,2,3,5-Tetrahydro-10-phenyl benzo[c] cyclopent[f] -[1,2] - diazepine $C_{18}H_{16}N_2$

2. 1,2,3,3a-Tetrahydro-10-phenyl benzo[c] cyclopent[f] -[1,2] - diazepine $C_{18}H_{16}N_2$

Stainless steel	1.00 m	2.1 mm
Permaphase[R] ODS	37-44 μm	50°C
40% methanol in water		1% octadecylsilyl chemically bonded
28 Bar	0.35 cm/sec	2 μl
Ultra-violet	254 nm	0.08 AUFS

Done, J.N., Knox, J.H., unpublished results

11.11   TRIPTYCENE DERIVATIVES

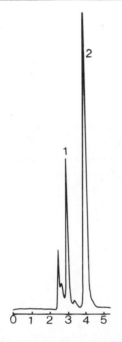

1. 9,10-Dimethyltriptycene $C_{22}H_{18}$
2. 9,10-Dimethoxytriptycene $C_{22}H_{18}O_2$

Stainless steel	1.00 m	2.1 mm
Zipax$^{(R)}$	20-37 μm	Ambient
Hexane		1% Polyethylene glycol 400
14 Bar		5 μl
Ultra-violet	254 nm	0.02 AUFS

Done, J.N., Knox, J.H., unpublished results.

11.12   CHLORINATED DIMETHOXYBENZENES

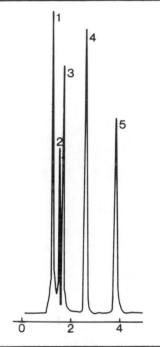

1. 2,3,5,6-Tetrachloro-1,4-dimethoxybenzene
   $C_8H_6Cl_4O_2$
2. 2,6-Dichloro-1,4-dimethoxybenzene
   $C_8H_8Cl_2O_2$
3. 2,3,5-Trichloro-1,4-dimethoxybenzene
   $C_8H_7Cl_3O_2$
4. 2,5-Dichloro-1,4-dimethoxybenzene
   $C_8H_8Cl_2O_2$
5. 2,3-Dichloro-1,4-dimethoxybenzene
   $C_8H_8Cl_2O_2$

Stainless steel	0.25 m	3.2 mm
Porous silica	5-6 μm	27°C
Hexane		30% BOP
41 Bar	1.0 cm^3/min	4 μl
Ultra-violet	254 nm	0.08 AUFS

Pore size of silica was about 350Å. Column packed using high pressure (6,000 psi) slurry packing procedure. BOP coating was added using the *in situ* coating method.

Kirkland, J.J., *Journal of Chromatographic Science,* **10,** 593 (1972). Reproduced by permission of Preston Technical Abstracts Co., Illinois.

11.13   NITRO-AROMATICS

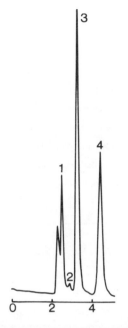

1. 3-Nitrofluoranthene $C_{16}H_9NO_2$
2. 2-Methoxy-7-nitrofluorene $C_{14}H_{11}NO_3$
3. 2,6-Dinitrotoluene $C_7H_6N_2O_4$
4. 2,4'-Dinitrobiphenyl $C_{12}H_8N_2O_4$

Glass	0.96 m	2.1 mm
Zipax$^{(R)}$	37-44 µm	Ambient
Cyclohexane		1% BOP
28 Bar	0.71 cm/sec	3 µl
Ultra-violet	254 nm	0.32 AUFS

Done, J.N., Knox, J.H., unpublished results

11.14   CHLORINATED BENZENES

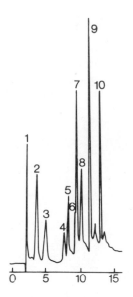

1. Benzene C_6H_6
2. Chlorobenzene C_6H_5Cl
3. *o*-Dichlorobenzene $C_6H_4Cl_2$
4. 1,2,3-Trichlorobenzene $C_6H_3Cl_3$
5. 1,3,5-Trichlorobenzene $C_6H_3Cl_3$
6. 1,2,4-Trichlorobenzene $C_6H_3Cl_3$
7. 1,2,3,4-Tetrachlorobenzene $C_6H_2Cl_4$
8. 1,2,4,5-Tetrachlorobenzene $C_6H_2Cl_4$
9. Pentachlorobenzene C_6HCl_5
10. Hexachlorobenzene C_6Cl_6

Stainless steel	1.00 m	2.1 mm
Permaphase$^{(R)}$ ODS	37-44 μm	60°C
Gradient: 40% methanol in water to 100% methanol at 8% per minute		1% octadecylsilyl chemically bonded
84 Bar		5 μl
Ultra-violet	254 nm	

11.15   ALKYL SUBSTITUTED ANTHRAQUINONES

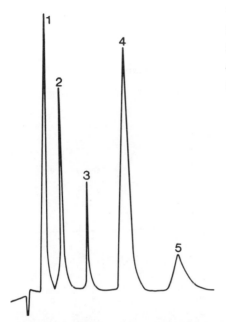

1. Anthraquinone $C_{14}H_8O_2$
2. 2-Methylanthraquinone $C_{15}H_{10}O_2$
3. 2-Ethylanthraquinone $C_{16}H_{12}O_2$
4. 1,4-Dimethylanthraquinone $C_{16}H_{12}O_2$
5. 2-t-Butylanthraquinone $C_{18}H_{16}O_2$

Stainless steel	1.00 m	2.1 mm
Permaphase$^{(R)}$ ODS	37-44 μm	
50% methanol in water		1% octadecylsilyl chemically bonded
Ultra-violet	254 nm	

Byrne, S.H., Schmit, J.A., Johnson, P.E., *Journal of Chromatographic Science,* **9,** 592 (1971). Reproduced by permission of Preston Technical Abstracts Co., Illinois.

11.16   CHLOROANTHRAQUINONES

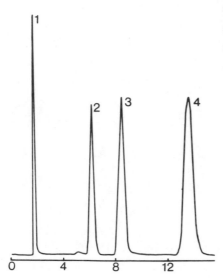

1. Toluene C_7H_8
2. 1-Chloroanthraquinone $C_{14}H_7ClO_2$
3. 1,2-Dichloroanthraquinone $C_{14}H_6Cl_2O_2$
4. 1,4-Dichloroanthraquinone $C_{14}H_6Cl_2O_2$

Glass	0.50m	2.1 mm
Spherisorb(R) A 20 Y	20 μm	Ambient
Hexane saturated with BOP		4% BOP
26 Bar	0.47 cm/sec	2.5 μl
Ultra-violet	254 nm	0.16 AUFS

The coating of BOP on the alumina is deactivating the surface so that the predominant mechanism of separation is adsorption.

Done, J.N., Knox, J.H., unpublished results.

11.17   AROMATIC ALCOHOLS

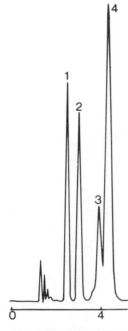

1. 2-Phenylpropan-2-ol $C_9H_{12}O$
2. α-Methylbenzyl alcohol $C_8H_{10}O$
3. Benzyl alcohol C_7H_8O
4. Cinnamyl alcohol $C_9H_{10}O$

Stainless steel	2.00 m	2 mm
Vydac(R) Adsorbent	30-44 μm	Ambient
1% amyl alcohol in iso-octane		None
193 Bar	3 cm^3/min	20 μl
Ultra-violet	254 nm	0.16 AUFS

Chromatronix Inc., Liquid Chromatography Application Number 14. Reproduced by permission of Chromatronix Inc., Berkeley.

11.18   PLASTICIZERS

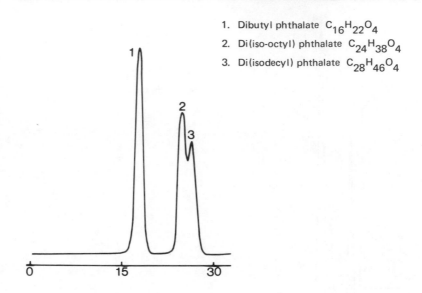

1. Dibutyl phthalate $C_{16}H_{22}O_4$
2. Di(iso-octyl) phthalate $C_{24}H_{38}O_4$
3. Di(isodecyl) phthalate $C_{28}H_{46}O_4$

	3.60 m	2.3 mm
Corasil(R) II	37-50 μm	Ambient
2% butyl acetate in iso-octane		None

	0.6 cm^3/min	
Ultra-violet	254 nm	

The sample was obtained by Gel Permeation Chromatography of a polymer containing plasticizers.

Separation by adsorption.

Reproduced by courtesy of Waters Associates, Inc.

11.19   PLASTICIZERS

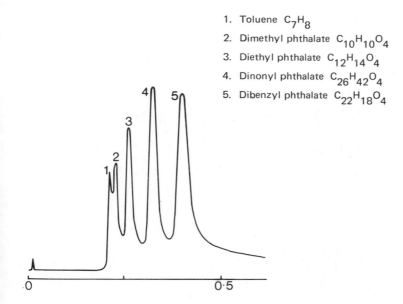

1. Toluene C_7H_8
2. Dimethyl phthalate $C_{10}H_{10}O_4$
3. Diethyl phthalate $C_{12}H_{14}O_4$
4. Dinonyl phthalate $C_{26}H_{42}O_4$
5. Dibenzyl phthalate $C_{22}H_{18}O_4$

	0.50 m	3.15 mm
Zipax[R]	20-37 μm	Ambient
Heptane		0.8% BOP
	4 cm/sec	

Separation by partition in contrast to the previous example.

Karger, B.L., Engelhardt, H., Conroe, K., Halasz, I., 'Evaluation of Porons Layer Beads in High Speed Liquid Chromatography' in *Gas Chromatography 1970* edited by Stock, R. and published by the Institute of Petroleum, London. Reproduced by permission of B.L. Karger.

11.20   ETHYLENE GLYCOLS

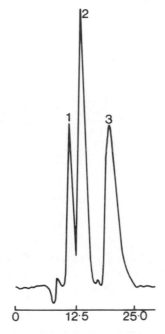

1. Ethylene glycol $C_2H_6O_2$
2. Diethylene glycol $C_4H_{10}O_3$
3. Triethylene glycol $C_6H_{14}O_4$

	0.30 m	3 mm
Porasil$^{(R)}$ 60	37-75 μm	Ambient
2.5% water in methylethyl ketone		None
	0.40 cm^3/min	
Refractive index		

11.21   TEST MIXTURE

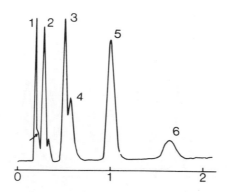

1. Butylbenzene $C_{10}H_{14}$
2. Quinoline C_9H_7N
3. Dimethyl phthalate $C_{10}H_{10}O_4$

4. Acetonylacetone $C_6H_{10}O_2$
5. Unknown
6. 2-Naphthylamine $C_{10}H_9N$

	0.087 m	3 mm
Chromosorb G	1-3 μm	Ambient
Iso-octane		None
	2.6 cm^3/min	1 μl
Ultra-violet	270 nm	

Note the use of a very small size porous material leading to quite high efficiency.

Ecker, E., *Chemiker-Zeitung*, **95**, 511 (1971). Reproduced by permission of Chemiker-Zeitung, Heidelberg, Redaktion.

11.22   TEST MIXTURE

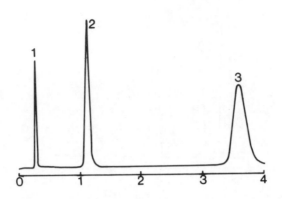

1. Benzene C_6H_6

2. Benzyl alcohol C_7H_8O

3. Benzanilide $C_{13}H_{11}NO$

Stainless steel	0.50 m	7.94 mm
Zipax$^{(R)}$	<37 μm	Ambient
Hexane		1% triethylene glycol
	3.3 cm/sec	
Ultra-violet	254 nm	

Large diameter column gives higher efficiency than smaller ones but at much increased cost. However sampler loading can also be increased. The 'infinite diameter' effect was also apparent with this column.

De Stefano, J.J., Beachell, H.C., *Journal of Chromatographic Science,* **8**, 434 (1970). Reproduced by permission of Preston Technical Abstract Co., Illinois.

11.23   POLAR ORGANIC MIXTURE

1. N-Methyl-p-toluenesulfonamide
   $C_8H_{11}NO_2S$
2. 3-Nitro-4-methoxybenzamide
   $C_8H_8N_2O_4$
3. N-Phenylcyanoacetamide $C_9H_8N_2O$
4. p-Nitrobenzenesulfonamide
   $C_6H_6N_2O_4S$

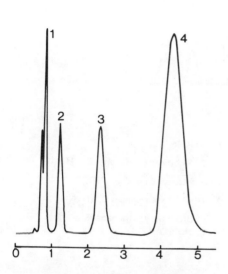

Stainless steel	1.00 m	2.1 mm
Zipax$^{(R)}$	<37 μm	27°C
50% isopropyl chloride in hexane		0.75% nitrile chemically bonded
95 Bar	4.35 cm^3/min	
Ultra-violet	254 nm	

This illustrates the use of a bonded stationary phase with a fairly polar mobile phase to separate a polar mixture.

Kirkland, J.J., De Stefano, J.J., *Journal of Chromatographic Science,* **8**, 309 (1970).
Reproduced by permission of Preston Technical Abstracts Co., Illinois.

11.24   TEST MIXTURE

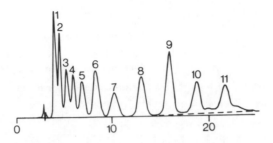

1. Tetrahydronaphthalene $C_{10}H_{12}$
2. Naphthalene $C_{10}H_8$
3. Acenaphthylene $C_{12}H_8$
4. Phenanthrene $C_{14}H_{10}$
5. Fluoranthene $C_{16}H_{10}$
6. 1-Methoxynaphthalene $C_{11}H_{10}O$

7. Anisole C_7H_8O
8. 2-Methoxynaphthalene $C_{11}H_{10}O$
9. o-Nitrotoluene $C_7H_7NO_2$
10. 1-Nitronaphthalene $C_{10}H_7NO_2$
11. m-Bromomethylbenzoate $C_8H_7BrO_2$

Glass	1.75 m	2.8 mm
Porasil$^{(R)}$ 60	35-75 μm	Ambient
Complex solvent programme		9% water
30 Bar		40 μl
Ultra-violet	260 nm	

This illustrates the use of a sophisticated solvent system to separate a mixture which has components of widely different polarity.

Synder, L.R., *Journal of Chromatographic Science*, **8**, 692 (1970). Reproduced by permission of Preston Technical Abstracts Co., Illinois.

APPENDIX

The following three tables give details of some of the liquid chromatography packing materials available at the present time. Table 1 lists most, but not all, of the pellicular materials available, all of which have been specially developed for use in HSLC. Table 2 lists the higher capacity porous packings available some of which were specially developed but others are merely smaller particle versions of adsorbents used for conventional column chromatography and TLC. Finally, Table 3 lists some of the large number of ion exchange resins produced along with a selection of the pellicular ion exchangers again specially developed for HSLC.

The Tables are by no means exhaustive and tend to show only those materials which were used in the separations illustrated in Part 2.

TABLE 1 Representative Pellicular Materials for HSLC

Type	Name	Description	Surface Area (m^2/g)	Use LLC	Use LSAC	Use LSPC	Supplier
Inactive Silica	Zipax(R)	25–37µm, spherical	1	X			Du Pont
Active Silica	Corasil(R)	37–50µm, spherical	7 & 14	X	X		Waters
	Vydac(R)	30–40µm, spherical	12	X	X		Chromatronix
	Pellosil(R)	37–44µm, spherical	4 & 8	X	X		Reeve Angel
	Perisorb(R)	30–40µm, spherical	10	X	X		E. Merck
Active Alumina	Pellumina(R)	37–44µm, spherical	4 & 8	X	X		Reeve Angel
Polymer Coated	Zipax(R) ANH	Cyanoethylsilicone coated on Zipax(R)	1			X	Du Pont
	Zipax(R) PAM	Polyamide coated on Zipax(R)	1			X	Du Pont
Chemically Bonded	Permaphase(R) ODS	Octadecylsilyl chemically bonded to Zipax(R)	<1			X	Du Pont
	Permaphase(R) ETH	Ether groups chemically bonded to Zipax(R)	<1			X	Du Pont
	Durapak(R) PEG 400	Carbowax 400 chemically bonded to Corasil(R)	7			X	Du Pont
	Bondapak(R) C_{18}/Corasil(R)	Octadecylsilyl chemically bonded to Corasil(R)	7			X	Waters
	Vydac(R) Reverse Phase	Octadecylsilyl chemically bonded to Vydac(R)	12			X	Chromatronix

LLC = Liquid-liquid partition, LSAC = Liquid solid adsorption, LSPC = Liquid solid partition

TABLE 2 Representative Porous Materials for HSLC

Type	Name	Description	Surface Area (m2/g)	Use LLC	LSAC	LSPC	Supplier
Silica	Porasil(R)	37–75µm, spherical, six types with different surface areas	up to 500	X	X		Waters Associates
	Sil-X(R)	36–45µm, irregular	300	X	X		Perkin-Elmer
	Merckosorb(R) SI 60	5,10,20 & 30 µm, irregular, Lichrosorb(R) in U.S.A.	500	X	X		E. Merck
	Zorbax-Sil (R)	5µm, spherical	300	X	X		Du Pont
	Spherisorb(R) S	5, 10 & 20µm, spherical	200	X	X		Phase Separations
Alumina	Spherisorb(R) A	5,10 & 20µm, spherical	95	X	X		Phase Separations
	Woelm alumina	18–32µm, irregular	200	X	X		M. Woelm
Chemically bonded	Durapak(R) Carbowax 400	Carbowax 400 chemically bonded to Porasil(R) C	50–100			X	Waters Associates
	Bondapak(R) C$_{18}$/Porasil(R)	Octadecylsilyl chemically bonded to Porasil(R) B	125–250			X	Waters Associates

LLC = Liquid-liquid partition, LSAC = Liquid-solid adsorption, LSPC = Liquid-solid partition

TABLE 3   Representative Ion Exchange packings for HSLC

Type	Name	Functional Group	Particle size (μm)	Capacity (μ equivalents/gm)	Supplier
Porous Anion	Durrum$^{(R)}$ DA-X4	$-\overset{+}{N}R_3$	20	2000	Durrum Chemical Co.
	Aminex$^{(R)}$ A-Series	$-\overset{+}{N}R_3$	9,13.5,17.5 & 20	3200	Bio-Rad Laboratories
	Bio-Rad$^{(R)}$ AG1-X2	$-\overset{+}{N}R_3$	37 & 37-74	3500	Bio-Rad Laboratories
Porous Cation	Hamilton HP-Series	$-SO_3H$	15 & 22	5200	Hamilton
	Durrum$^{(R)}$ DC-A Series	$-SO_3H$	8,12 & 18	5000	Durrum Chemical Co.
	Aminex$^{(R)}$ A-Series	$-SO_3H$	9,13,17.5 & 20	5000	Bio-Rad Laboratories
Pellicular Anion	Zipax$^{(R)}$ SAX	$-\overset{+}{N}R_3$	25-37	10	Du Pont
	Zipax$^{(R)}$ AAX —	$-\overset{+}{N}R_3$	25-37	12	Du Pont
	Pellicular Anion	$-\overset{+}{N}R_3$	40	12	Varian
	AS-Pellionex-SAX$^{(R)}$	$-\overset{+}{N}R_3$	44-53	10	Reeve Angel
Pellicular Cation	Zipax$^{(R)}$ SCX	$-SO_3H$	25-37	3.2	Du Pont
	Pellicular Cation	$-SO_3H$	40	10	Varian
	HS-Pellionex SCX$^{(R)}$	$-SO_3H$	44-53	8-10	Reeve Angel

AUTHOR INDEX

(Page numbers are given for names cited in Part 1; details of each reference are given on page 39. Chromatogram numbers are given for names cited in Part 2; details of each reference are given at the foot of each chromatogram.)

Part 1

194

Part 2

FORMULA INDEX

(The numbers refer to the chromatograms)

C_2

$C_2H_4O_2$	Acetic acid		2.08
$C_2H_6ClO_2PS$	*O,O*-Dimethyl chlorothiophosphate		6.14
$C_2H_6O_2$	Ethylene glycol		11.20

C_3

$C_3H_4O_3$	Pyruvic acid		2.08
C_3H_6O	Acetone	2.04	8.05
$C_3H_6O_3$	Lactic acid		2.08
C_3H_7NOS	*S*-Methyl *N*-hydroxythioacetimidate		6.13

C_4

$C_4H_4N_2O_2$	Uracil		7.05
$C_4H_4O_4$	Fumaric acid	2.01	2.08
	Maleic acid		2.01
$C_4H_5N_3O$	Cytosine		7.05
$C_4H_6O_4$	Succinic acid		2.08
$C_4H_6O_5$	Malic acid		2.07
$C_4H_8N_2O_2S$	*S*-Methyl *N*-(carbamoyloxy)thioacetimidate		6.12
$C_4H_8O_3$	3-Hydroxybutyric acid		2.08
$C_4H_{10}O_3$	Diethylene glycol		11.20

C_5

C_5HCl_4NO	3,4,5,6-Tetrachloropyridin-2-ol	2.02
$C_5H_2Cl_3NO$	3,4,5-Trichloropyridin-2-ol	2.02
	3,5,6-Trichloropyridin-2-ol	2.02
$C_5H_3Cl_2NO$	3,5-Dichloropyridin-2-ol	2.02
$C_5H_4N_4O$	Hypoxanthine	7.05
C_5H_5N	Pyridine	3.09
$C_5H_5N_5$	Adenine	7.05
$C_5H_5N_5O$	Guanine	7.05

$C_5H_6O_5$	2-Oxoglutaric acid				2.07	2.08
$C_5H_8O_4$	Glutaric acid					2.08
$C_5H_{10}N_2O_2S$	Methomyl (S-Methyl N-(methylcarbamoyloxy)thioacetimidate)				6.12	6.13

$$C_6$$

C_6Cl_6	Hexachlorobenzene					11.14
C_6HCl_5	Pentachlorobenzene					11.14
$C_6H_2Cl_4$	1,2,3,4-Tetrachlorobenzene					11.14
	1,2,4,5-Tetrachlorobenzene					11.14
$C_6H_3Cl_3$	1,2,3-Trichlorobenzene					11.14
	1,2,4-Trichlorobenzene					11.14
	1,3,5-Trichlorobenzene					11.14
$C_6H_3Cl_3O$	2,4,6-Trichlorophenol					4.05
$C_6H_4Cl_2$	o-Dichlorobenzene					11.14
$C_6H_4Cl_2O$	2,4-Dichlorophenol					4.05
	2,6-Dichlorophenol					4.05
$C_6H_4N_2$	Isonicotinonitrile					3.14
	Nicotinonitrile				3.13	3.14
	Picolinonitrile					3.14
$C_6H_5BrO_2$	Bromohydroquinone					4.07
C_6H_5Cl	Chlorobenzene					11.14
C_6H_5ClO	o-Chlorophenol					4.05
	p-Chlorophenol					4.05
$C_6H_5ClO_3S$	p-Chlorobenzene sulfonic acid					5.02
$C_6H_5NO_2$	Pyridine-2-carboxylic acid					2.03
	Pyridine-3-carboxylic acid					2.03
	Pyridine-4-carboxylic acid					2.03
$C_6H_5NO_3$	o-Nitrophenol					4.06
	m-Nitrophenol					4.06
	p-Nitrophenol				4.06	6.04
C_6H_6	Benzene	1.02	1.03	1.04	1.06
		1.07	3.05	11.14	11.22
$C_6H_6Cl_6$	Lindane$^{(R)}$ (γ-1,2,3,4,5,6-Hexachlorocyclohexane)				6.10	6.14
$C_6H_6N_2O$	Isonicotinamide					3.13

$C_6H_6N_2O$	Nicotinamide			3.13	3.14
	Picolinamide			2.03	3.13
$C_6H_6N_2O_2$	o-Nitroaniline				3.08
	m-Nitroaniline				3.08
	p-Nitroaniline				3.08
$C_6H_6N_2O_4S$	p-Nitrobenzenesulfonamide				11.23
C_6H_6O	Phenol			4.02	4.04
$C_6H_6O_2$	Hydroquinone			4.06	4.07
	Pyrocatechol			4.06	4.07
	Resorcinol			4.06	4.07
$C_6H_6O_3S$	Benzenesulfonic acid			5.02	5.07
$C_6H_6O_4S$	4-Hydroxybenzenesulfonic acid				5.07
$C_6H_6O_6$	cis-Propene-1,2,3-tricarboxylic acid				2.07
C_6H_7N	Aniline			3.03	3.05
$C_6H_7NO_2$	6-Methylpyridin-2,4-diol				2.09
$C_6H_7NO_3S$	p-Anilinesulfonic acid				5.06
	Orthanilic acid				5.03
$C_6H_7N_5$	6-Methylaminopurine				7.05
$C_6H_8N_2$	o-Phenylenediamine				3.07
	m-Phenylenediamine				3.07
	p-Phenylenediamine				3.07
$C_6H_8O_6$	Vitamin C (Ascorbic acid)				8.07
$C_6H_8O_7$	Citric acid				2.07
	Isocitric acid				2.07
$C_6H_{10}O_2$	Acetonyl acetone				11.21
$C_6H_{12}O_2$	Hexanoic acid				2.05
$C_6H_{14}O_4$	Triethylene glycol				11.20

C_7

$C_7H_4Cl_2N_2$	5,6-Dichlorobenzimidazole				11.05
$C_7H_6Cl_3NO$	2,3,5-Trichloro-6-ethoxypyridine				2.02
$C_7H_6N_2O_4$	2,6-Dinitrotoluene				11.13
$C_7H_6O_2$	Benzoic acid	2.04	2.06	8.11	8.15
$C_7H_7Cl_2NO$	Clopidol (3,5-Dichloro-2,6-dimethylpyridin-4-ol)				8.08

$C_7H_7NO_2$	o-Nitrotoluene				11.24
	Salicylamide				8.10
$C_7H_7N_3$	2-Aminobenzimidazole				11.05
C_7H_8	Toluene	1.09	3.04	11.16	11.19
$C_7H_8N_4O_2$	Theobromine				10.10
	Theophylline			8.15	10.10
C_7H_8O	Anisole				11.24
	Benzyl alcohol			11.17	11.22
	o-Cresol				4.04
	m-Cresol			4.02	4.04
	p-Cresol				4.02
$C_7H_8O_3S$	p-Toluenesulfonic acid				5.02
C_7H_9N	o-Toluidine				3.01
C_7H_9NO	2,6-Dimethylpyridin-4-ol				2.09
$C_7H_{13}N_3O_3S$	S-Methyl (dimethylcarbamoyl)-N-(methylcarbamoyloxy)thiocarbamate				6.12

C_8

$C_8H_6Cl_2O_3$	2,4-Dichlorophenoxyacetic acid				6.07
$C_8H_6Cl_4O_2$	2,3,5,6-Tetrachloro-1,4-dimethoxybenzene				11.12
$C_8H_6O_4$	Isophthalic acid				2.04
	Phthalic acid				2.04
	Terephthalic acid			2.04	2.06
$C_8H_7BrO_2$	m-Bromomethylbenzoate				11.24
$C_8H_7Cl_3O_2$	2,3,5-Trichloro-1,4-dimethoxybenzene				11.12
C_8H_8	Styrene	1.04	1.07		1.09
$C_8H_8Cl_2O_2$	2,3-Dichloro-1,4-dimethoxybenzene				11.12
	2,5-Dichloro-1,4-dimethoxybenzene				11.12
	2,6-Dichloro-1,4-dimethoxybenzene				11.12
$C_8H_8N_2O_4$	4-Methoxy-3-nitrobenzamide				11.23
$C_8H_8O_2$	o-Toluic acid				2.06
$C_8H_8O_4$	Dehydroacetic acid (3-Acetyl-6-methyl-2H-pyran-2,4(3H)-dione)				2.09
$C_8H_9NO_2$	Paracetamol (p-Hydroxyacetanilide)				8.10

$C_8H_9NO_3$	4-Hydroxy-2,6-dimethylpyridine-3-carboxylic acid			2.09
$C_8H_9NO_5S$	Acetanilide sulfate (p-(Acetylamino)benzenesulfonic acid)			8.17
C_8H_{10}	m-Xylene			3.03
$C_8H_{10}NO_5PS$	Methyl parathion (O-O-Dimethyl O-p-nitrophenyl phosphorothionate)	6.03	6.04	6.14
$C_8H_{10}N_4O_2$	Caffeine	8.10	8.11	10.10
$C_8H_{10}O$	α-Methylbenzyl alcohol			11.17
	2,3-Xylenol			4.02
	2,4-Xylenol			4.04
	2,6-Xylenol		4.02	4.03
	3,4-Xylenol			4.02
	3,5-Xylenol			4.02
$C_8H_{10}O_3S$	2,5-Dimethylbenzenesulfonic acid			5.02
$C_8H_{11}N$	N,N-Dimethylaniline		3.01	3.04
	2,3-Dimethylaniline			3.03
	2,5-Dimethylaniline			3.05
	2,6-Dimethylaniline	3.03	3.04	3.05
	N-Ethylaniline			3.06
$C_8H_{11}NO_2S$	N,N-Dimethylbenzenesulfonamide		8.12	11.23
	N-Methyl-p-toluenesulfonamide		8.12	11.23
$C_8H_{11}NO_3S$	4,6-Dimethylorthanilic acid			5.03
$C_8H_{12}S_3$	5-t-Butyl-4-methyl-1,2-dithiole-3-thione			11.06
	4-Neopentyl-1,2-dithiole-3-thione			11.06

C_9

$C_9H_4Cl_3NO_2S$	Folpet (N-Trichloromethylthiophthalimide)			6.02
$C_9H_6CrO_3$	Benzenetricarbonylchromium (Tricarbonyl(η-benzene)chromium)			11.04
$C_9H_6O_2$	Coumarin			10.07
C_9H_7N	Isoquinoline		3.01	3.09
	Quinoline	3.01	3.09	11.21
C_9H_7NO	8-Quinolinol			3.09
C_9H_8	Indene		1.04	1.07
$C_9H_8N_2O$	N-Phenylcyanoacetamide			11.23

$C_9H_8O_2$	3,4-Dihydrocoumarin					10.07
$C_9H_8O_4$	Aspirin (O-Acetylsalicylic acid)				8.10	8.11
C_9H_{10}	Indane					1.05
$C_9H_{10}Cl_2N_2O$	Diuron (1-(3,4-Dichlorophenyl)-3,3-dimethylurea)					6.01
$C_9H_{10}Cl_2N_2O$	Linuron (1-(3,4-Dichlorophenyl)-3-methoxy-3-methylurea)					6.01
$C_9H_{10}N_2$	5,6-Dimethylbenzimidazole					11.05
$C_9H_{10}O$	Cinnamyl alcohol					11.17
$C_9H_{11}ClN_2O$	Monuron (1-p-Chlorophenyl-3,3-dimethylurea)					6.01
$C_9H_{11}N_3$	2-Amino-5,6-dimethylbenzimidazole					11.05
C_9H_{12}	Mesitylene					3.05
$C_9H_{12}N_2O$	Fenuron (1,1-Dimethyl-3-phenylurea)					6.01
$C_9H_{12}O$	2-Phenylpropan-2-ol					11.17
$C_9H_{13}NO_2S$	N,N-Dimethyl-p-toluenesulfonamide					8.12
$C_9H_{13}N_2O_9P$	Uridine-2'-monophosphoric acid				7.02	7.04
	Uridine-3'-monophosphoric acid				7.02	7.04
	Uridine-5'-monophosphoric acid				7.01	7.07
$C_9H_{14}N_2O_{12}P_2$	Uridine-5'-diphosphoric acid					7.01
$C_9H_{14}N_3O_8P$	Cytidine-2'-monophosphoric acid				7.02	7.04
	Cytidine-3'-monophosphoric acid				7.02	7.04
	Cytidine-5'-monophosphoric acid				7.01	7.07
$C_9H_{15}NO_4S$	Dexedrine ((+)-α-Methylphenethylammonium sulfate)					8.14
$C_9H_{15}N_2O_{15}P_3$	Uridine-5'-triphosphoric acid					7.01
$C_9H_{15}N_3O_{11}P_2$	Cytidine-5'-diphosphoric acid					7.01
$C_9H_{16}N_3O_{14}P_3$	Cytidine-5'-triphosphoric acid					7.01

C_{10}

$C_{10}H_5Cl_7$	Heptachlor (1,4,5,6,7,8,8-Heptachloro-3a,4,7,7a-tetrahydro-4,7-methanoindene)						6.14
$C_{10}H_7NO_2$	1-Nitronaphthalene						11.24
$C_{10}H_8$	Azulene						1.02
	Naphthalene	1.02	1.04	1.05	1.06		
			1.07	1.09			11.24
$C_{10}H_8CrO_3$	Toluenetricarbonylchromium (Tricarbonyl(η-toluene)chromium)						11.04

$C_{10}H_8N_2O_6S$	3-Carboxy-5-hydroxy-1-p-sulfophenylpyrazole			5.06	
$C_{10}H_8O$	1-Naphthol		4.06	6.08	
	2-Naphthol			4.06	
$C_{10}H_8O_2$	6-Methylcoumarin			10.07	
	Naphthalene-1,5-diol			4.07	
$C_{10}H_8O_3S$	Naphthalene-2-sulfonic acid			5.02	
$C_{10}H_8O_4$	7-Hydroxy-6-methoxycoumarin			10.07	
$C_{10}H_8O_4S$	Schaeffer's salt (6-Hydroxy-2-naphthalenesulfonic acid)			5.04	
$C_{10}H_8O_7S_2$	G salt (7-Hydroxy-1,3-naphthalenedisulfonic acid)			5.04	
	R salt (3-Hydroxy-2,7-naphthalenedisulfonic acid)			5.04	
$C_{10}H_9N$	1-Naphthylamine			3.01	
	2-Naphthylamine			3.01	11.21
$C_{10}H_{10}O_4$	Dimethyl phthalate			11.19	11.21
$C_{10}H_{12}$	1,2,3,4-Tetrahydronaphthalene		1.04	1.07	11.24
$C_{10}H_{12}N_2O$	5,6-Dimethyl-2-benzimidazolemethanol			11.05	
$C_{10}H_{12}N_5O_6P$	3′,5′-Cyclic adenosine monophosphate			7.03	7.07
$C_{10}H_{13}ClN_2O_3S$	Chloropropamide (N-p-Chlorobenzenesulfonyl-N′-propylurea)			8.02	
$C_{10}H_{13}NO_2$	N-Methyl-3,4-(methylenedioxy)phenethylamine			8.14	
	Phenacetin (N-Acetyl-p-phenetidine)			8.10	8.11
$C_{10}H_{13}N_4O_8P$	Inosine-5′-monophosphoric acid			7.03	7.07
$C_{10}H_{14}$	Butylbenzene			11.21	
$C_{10}H_{14}NO_5PS$	Parathion (O,O-Diethyl O-p-nitrophenyl phosphorothionate)		6.02	6.03	6.04
$C_{10}H_{14}N_5O_7P$	Adenosine-2′-monophosphoric acid			7.02	
	Adenosine-3′-monophosphoric acid			7.02	7.04
	Adenosine-5′-monophosphoric acid	7.01	7.03	7.07	7.08
$C_{10}H_{14}N_5O_8P$	Guanosine-2′-monophosphoric acid			7.02	
	Guanosine-3′-monophosphoric acid			7.02	7.04
	Guanosine-5′-monophosphoric acid			7.01	7.07
$C_{10}H_{14}O$	2-t-Butylphenol			4.03	4.08
$C_{10}H_{14}O_2$	t-Butylhydroquinone			4.07	

$C_{10}H_{15}N$	N,N-Diethylaniline		3.06
$C_{10}H_{15}NO$	Ephedrine (2-Methylamino-1-phenylpropan-1-ol)	8.14	8.15
$C_{10}H_{15}N_5O_{10}P_2$	Adenosine-5′-diphosphoric acid		7.01
$C_{10}H_{15}N_5O_{11}P_2$	Guanosine-5′-diphosphoric acid		7.01
$C_{10}H_{15}O_2PS_2$	Dyfonate[R] (O,O-Diethyl S-phenyl phosphorothioate)		6.06
$C_{10}H_{16}N_5O_{13}P_3$	Adenosine-5′-triphosphoric acid		7.01
$C_{10}H_{16}N_5O_{14}P_3$	Guanosine-5′-triphosphoric acid		7.01
$C_{10}H_{18}$	Decalin		1.02

C_{11}

$C_{11}H_8O_2$	Vitamin K_3 (2-Methyl-1,4-naphthoquinone)	8.07
$C_{11}H_{10}CrO_3$	m-Xylenetricarbonylchromium (Tricarbonyl(η-m-xylene)chromium)	11.04
$C_{11}H_{10}O$	1-Methoxynaphthalene	11.24
	2-Methoxynaphthalene	11.24
$C_{11}H_{12}Cl_2O_3$	2,4-Dichlorophenoxyacetic acid, isopropyl ester (Isopropyl 2,4-dichlorophenoxyacetate)	6.07
$C_{11}H_{12}NO_4PS_2$	Imidan (N-[Dimethoxy(thiophosphinyl)thiomethyl] phthalimide)	6.02
$C_{11}H_{16}O$	6-t-Butyl-o-cresol	4.03
	6-t-Butyl-m-cresol	4.08
	6-t-Butyl-p-cresol	4.08
$C_{11}H_{18}N_2O_3$	Amobarbital (5-Ethyl-5-isopentylbarbituric acid)	8.06
$C_{11}H_{18}N_2O_4$	Hydroxyamobarbital (5-Ethyl-5-(3-hydroxy-3-methylbutyl) barbituric acid)	8.06

C_{12}

$C_{12}H_8$	Acenaphthylene	11.24
$C_{12}H_8ClNS$	2-Chlorophenothiazine	11.09
	3-Chlorophenothiazine	11.09
$C_{12}H_8ClNO_2S$	p-Chlorophenyl o-nitrophenyl sulfide	11.09

Formula	Name				
$C_{12}H_8Cl_6$	Aldrin (1,2,3,4,10,10-Hexachloro-1,4,4a,5,8, 8a-hexahydro-1,4-*endo-exo*-5, 8-dimethanonaphthalene)			6.10	6.14
$C_{12}H_8Cl_6O$	Endrin (1,2,3,4,10,10-Hexachloro-6,7-epoxy-1, 4,4a,5,6,7,8,8a-octahydro-1,4-*endo-endo*-5, 8-dimethanonaphthalene)			6.10	6.14
$C_{12}H_8N_2$	Phenazine				3.10
$C_{12}H_8N_2O_4$	2,4'-Dinitrobiphenyl				11.13
$C_{12}H_9N$	Carbazole			1.05	3.01
$C_{12}H_{10}$	Acenaphthene				1.05
	Biphenyl	1.03	1.06	1.07	1.09
$C_{12}H_{10}N_2$	Azobenzene			3.01	3.12
$C_{12}H_{10}O_3S$	Biphenyl-4-sulfonic acid				5.01
$C_{12}H_{10}O_4S$	4'-Hydroxybiphenyl-4-sulfonic acid				5.01
$C_{12}H_{10}O_6S_2$	Biphenyl-4,4'-disulfonic acid				5.01
$C_{12}H_{11}N$	Diphenylamine				3.06
$C_{12}H_{11}NO_2$	Carbaryl (1-Naphthyl-*N*-methylcarbamate)				6.08
$C_{12}H_{11}N_3$	4-Aminoazobenzene				3.12
$C_{12}H_{12}CrO_3$	Mesitylenetricarbonylchromium (Tricarbonyl(η-mesitylene)chromium)				11.04
$C_{12}H_{12}N_2O_2S$	4,4'Diaminodiphenyl sulfone (Bis-(*p*-aminophenyl)sulfone)				8.04
$C_{12}H_{12}N_2O_3$	Phenobarbital (5-Ethyl-5-phenylbarbituric acid)		8.06	8.11	8.15
$C_{12}H_{12}N_2O_4$	Hydroxyphenobarbital (5-Ethyl-5-*p*-hydroxyphenylbarbituric acid)				8.06
$C_{12}H_{12}O_3$	7-Ethoxy-4-methylcoumarin				10.07
$C_{12}H_{14}Cl_2O_3$	2,4-Dichlorophenoxyacetic acid, isobutyl ester (Isobutyl 2,4-dichlorophenoxyacetate)				6.07
$C_{12}H_{14}N_2O_4$	Ketohexobarbital (5-(3-oxo-1-cyclohexenyl)-1, 5-dimethyl barbituric acid)				8.06
$C_{12}H_{14}O_4$	Diethyl phthalate				11.19
$C_{12}H_{15}N_5O_3$	9-(4-*c*-Cyclopropyl-α-D-*ribo*-tetrafuranosyl) adenine				7.06

$C_{12}H_{15}N_5O_3$	9-(4-c-Cyclopropyl-β-D-*ribo*-tetrafuranosyl) adenine				7.06
$C_{12}H_{18}N_2O_3S$	Tolbutamide (1-Butyl-3-[(toluene-*p*-sulfonyl) carbamoyl] urea)				8.02
$C_{12}H_{18}O$	2-t-Butyl-4,6-dimethylphenol				4.08
$C_{12}H_{19}NO_2$	STP (α-Methyl-2,5-dimethoxy-4- methylphenethylamine)				8.14
$C_{12}H_{24}O_2$	Dodecanoic acid				2.05

C_{13}

$C_{13}H_7Br_4NO_2$	2',3,4',5-Tetrabromosalicylanilide				8.05
$C_{13}H_8Br_3NO_2$	3,4',5-Tribromosalicylanilide				8.05
$C_{13}H_9Br_2NO_2$	4',5-Dibromosalicylanilide				8.05
$C_{13}H_9N$	Acridine			3.10	3.11
	Benzo[c]quinoline			3.10	3.11
	Benzo[f]quinoline			3.10	3.11
	Benzo[h]quinoline		3.01	3.10	3.11
$C_{13}H_{10}$	Fluorene			1.05	1.07
$C_{13}H_{10}BrNO_2$	4'-Bromosalicylanilide				8.05
$C_{13}H_{11}NO$	Benzanilide			6.13	11.22
$C_{13}H_{13}NO_2S$	*N*-Phenyl-*p*-toluenesulfonamide				8.12
$C_{13}H_{14}N_2$	Di-(*o*-aminophenyl)methane				3.02
	Di-(*m*-aminophenyl)methane				3.02
	Di-(*p*-aminophenyl)methane				3.02

C_{14}

$C_{14}H_6Cl_2O_2$	1,2-Dichloroanthraquinone				11.16
	1,4-Dichloroanthraquinone				11.16
$C_{14}H_7ClO_2$	1-Chloroanthraquinone				11.16
$C_{14}H_8O_2$	Anthraquinone				11.15
$C_{14}H_9Cl_5$	o,p'-DDT (1,1,1-Trichloro-2-(o-chlorophenyl)- 2-(p -chlorophenyl)ethane)				6.14
	p,p'-DDT (1,1,1-Trichloro-2, 2-bis(p-chlorophenyl)ethane)			6.10	6.14
$C_{14}H_{10}$	Anthracene		1.06	1.07	1.09
	Phenanthrene	1.05	1.06	1.07	11.24

$C_{14}H_{10}Cl_4$ — p,p'-DDD (1,1-Dichloro-2,
2-bis(p-chlorophenyl)ethane)   6.10 6.14

$C_{14}H_{11}NO_3$ — 2-Methoxy-7-nitrofluoranthene   11.13

$C_{14}H_{12}O_3$ — 2-Hydroxy-4-methoxybenzophenone   4.09

$C_{14}H_{14}N_2O_3S$ — 4-Acetamido-4'-aminodiphenyl sulfone
(p-Acetamidophenyl-p-amino phenyl
sulfone)   8.04

$C_{14}H_{15}NO_4PS$ — EPN (O-Ethyl (O-p-nitrophenyl)
phenylphosphorothionate)   6.14

$C_{14}H_{21}N_3O_3$ — Tandex$^{(R)}$(1,1-Dimethyl-3-[m-
(t-butylcarbamoyloxy)phenyl] urea)   6.11

$C_{14}H_{21}N_3O_3S$ — Tolazamide (1-Azepinyl-3-benzenesulfonylurea)   8.02

$C_{14}H_{22}O_2$ — 2,5-Di-t-butylhydroquinone   4.07

$C_{14}H_{28}O_2$ — Tetradecanoic acid   2.05

C_{15}

$C_{15}H_{10}O_2$ — 2-Methylanthraquinone   11.15

$C_{15}H_{11}ClN_2O$ — N-Demethylated metabolite of Diazepam
(7-Chloro-1,2-dihydro-2-oxo-5-phenyl-3H-1,
4-benzodiazepine)   8.09

$C_{15}H_{11}ClN_2O_2$ — Oxazepam (7-Chloro-1,
2-dihydro-3-hydroxy-2-oxo-5-phenyl-3H-1,
4-benzodiazepine)   8.09

$C_{15}H_{12}Br_4O_2$ — 2,2',6,6'-Tetrabromobisphenol A
(2,2',6,6'-Tetrabromo-4,
4'-isopropylidenediphenol)   4.01

$C_{15}H_{12}CrO_3$ — Tricarbonyl(η-1,2,3,4,4a,8a-2,
3-dimethylnaphthalene)chromium   11.03
Tricarbonyl(η-4a,5,6,7,8,8a-2,
3-dimethylnaphthalene)chromium   11.03

$C_{15}H_{12}N_2O_2$ — Diphenylhydantoin   8.03

$C_{15}H_{12}N_2O_3$ — 5-p-Hydroxyphenyl-5-phenylhydantoin   8.03

$C_{15}H_{13}Br_3O_2$ — 2,2',6-Tribromobisphenol A
(2,2',6-Tribromo-4,4'-isopropylidenediphenol)   4.01

$C_{15}H_{14}Br_2O_2$ — 2,2'-Dibromobisphenol A
(2,2'-Dibromo-4,4'-isopropylidenediphenol)   4.01

$C_{15}H_{14}Br_2O_2$	2,6-Dibromobisphenol A (2,6-Dibromo-4, 4'-isopropylidenediphenol)	4.01
$C_{15}H_{15}BrO_2$	2-Bromobisphenol A (2-Bromo-4, 4'-isopropylidenediphenol)	4.01
$C_{15}H_{16}O_2$	Bisphenol A (4,4'-Isopropylidenediphenol)	4.01
$C_{15}H_{20}N_2O_4S$	Acetohexamide (1-(p-Acetylbenzenesulfonyl)-3-phenylurea)	8.02
$C_{15}H_{24}O$	2,6-Di-t-butyl-p-cresol	4.09

C_{16}

$C_{16}H_9NO_2$	2-Nitrofluoranthene		11.13
$C_{16}H_{10}$	Fluoranthene	1.06	11.24
	Pyrene	1.05	1.06
$C_{16}H_{12}N_4O_9S_2$	FD&C Yellow 5 (3-Carboxy-5-hydroxy-1-p-sulfophenyl-4-p-sulfophenylazopyrazole)		5.06
$C_{16}H_{12}O_2$	1,4-Dimethylanthraquinone		11.15
	2-Ethylanthraquinone		11.15
$C_{16}H_{13}ClN_2O$	Diazepam (7-Chloro-1,2-dihydro-1-methyl-2-oxo-5-phenyl-3H-1,4-benzodiazepine)		8.09
$C_{16}H_{13}ClN_2O_2$	Hydroxylated metabolite of Diazepam (7-Chloro-1,2-dihydro-3-hydroxy-1-methyl-2-oxo-5-phenyl-3H-1,4-benzodiazepine)		8.09
$C_{16}H_{13}N$	N-Phenyl-2-naphthylamine		3.06
$C_{16}H_{15}Cl_3O_2$	Methoxychlor (1,1,1-Trichloro-2,2-bis-(p-methoxyphenyl)ethane)		6.09
$C_{16}H_{16}ClN_3O_3S$	Metolazone (7-Chloro-1,2,3,4-tetrahydro-2-methyl-4-oxo-3-o-tolyl-6-quinazolinesulfonamide)		8.20
$C_{16}H_{16}Cl_2O_2$	1,1-Dichloro-2,2-di(p-methoxyphenyl)ethane		6.09
$C_{16}H_{16}N_2O_4S$	4,4'-Diacetamidodiphenyl sulfone (Bis-(p-acetamidophenyl)sulfone)		8.04
$C_{16}H_{17}NO_8$	Acetanilide glucuronide		8.17
$C_{16}H_{18}BrN_3$	3'-Bromo-4-diethylaminoazobenzene		3.12
$C_{16}H_{18}N_4O_2$	4-Diethylamino-3'-nitroazobenzene		3.12

$C_{16}H_{18}N_4O_2$	4-Diethylamino-4'-nitroazobenzene			3.12
$C_{16}H_{19}N_3$	4-Diethylaminoazobenzene			3.12
$C_{16}H_{20}O_6P_2S_3$	Abate[R] (*O,O,O',O'*-Tetramethyl *O,O'*-thiodi-*p*-phenylene phosphorodithioate)			6.05
$C_{16}H_{22}Cl_2O_3$	2,4-Dichlorophenoxyacetic acid, ethylhexyl ester (Ethylhexyl 2,4-dichlorophenoxyacetate)			6.07
$C_{16}H_{22}O_4$	Dibutyl phthalate			11.18
$C_{16}H_{32}O_2$	Palmitic acid			2.05

C_{17}

$C_{17}H_{12}$	11*H*-Benzo[*b*] fluorene			1.01
$C_{17}H_{12}O_6$	Aflatoxin B_1			10.03
$C_{17}H_{12}O_7$	Aflatoxin G_1			10.03
$C_{17}H_{12}S_2Se$	2,5-Diphenyl-6a-selenathiophthen			11.08
$C_{17}H_{12}S_3$	2,5-Diphenyl-6a-thiathiophthen			11.08
$C_{17}H_{14}O_6$	Aflatoxin B_2			10.03
$C_{17}H_{14}O_7$	Aflatoxin G_2			10.03
$C_{17}H_{19}NO_3$	Morphine			10.11
$C_{17}H_{20}N_2O_4S$	Methyl ester of Benzylpenicillin			8.18
$C_{17}H_{20}N_4O_6$	Riboflavin (Vitamin B_2)		7.08	8.07
$C_{17}H_{21}N$	Benzphetamine (*N*-(+)-Benzyl-*N*-α-dimethylphenethylamine)			8.14
	α-Benzyl-3,4-dimethylphenethylamine			8.14
$C_{17}H_{21}N_4O_9P$	Flavin mononucleotide			7.08

C_{18}

$C_{18}H_{12}$	Chrysene		1.05	1.06
$C_{18}H_{14}$	*o*-Terphenyl			1.11
	m-Terphenyl	1.03	1.09	1.11
	p-Terphenyl			1.11
$C_{18}H_{16}N_2$	1,2,3,3a-Tetrahydro-10-phenylbenzo[*c*] cyclopent[*f*]-[1,2]-diazepine			11.10

C_{19}

$C_{19}H_{26}O_3$	4-Androstene-11β-ol-3,17-dione			
	(11β-Hydroxyandrosta-1,4-dien-3-one)			9.06
$C_{19}H_{28}O_2$	Testosterone (17β-Hydroxyandrost-			
	4-en-3-one)	9.05	9.06	9.07
$C_{19}H_{30}O_2$	Androsterone			9.07

C_{20}

$C_{20}H_{12}$	Benz[a]pyrene			1.06
	Benz[e]pyrene			1.06
$C_{20}H_{21}NO_4$	Papaverine			10.11
$C_{20}H_{30}O$	Vitamin A	8.07	8.13	8.19

C_{21}

$C_{21}H_{14}$	13H-Dibenzo[a.g]fluorene		1.01
$C_{21}H_{14}N_2O_3$	N-Benzoyl-1,5-diaminoanthraquinone		5.05
$C_{21}H_{18}O_3$	3,4-Dibenzyloxybenzaldehyde		11.07
$C_{21}H_{22}N_2O_2$	Strychnine		10.05
$C_{21}H_{23}NO_5$	Cryptopine		10.11
$C_{21}H_{26}O_5$	Prednisone (17α,21-Dihydroxypregna-1,		
	4-diene-3,11,20-trione)		9.10
$C_{21}H_{28}O_4$	11-Dehydrocorticosterone		
	(21-Hydroxypregn-4-ene-3,11,20-trione)		9.02
	4-Pregnene-17α-ol-3,11,20-trione		
	(17α-Hydroxypregn-4-ene-3,11,20-trione)		9.03
$C_{21}H_{28}O_5$	Aldosterone		9.02
	Cortisone		9.02
	Prednisolone (11β,17α,21-Trihydroxypregna-1,		
	4-dien-3,20-dione)	9.08	9.10
$C_{21}H_{28}O_6$	6β-Hydroxycortisone	9.02	9.03
$C_{21}H_{30}O_2$	Progesterone	9.05	9.07
$C_{21}H_{30}O_3$	Deoxycorticosterone		9.02
$C_{21}H_{30}O_4$	Corticosterone		9.02
	11-Deoxycortisol		9.02
	4-Pregnene-11β,17α-diol-3,20-dione (11β,		
	17α-Dihydroxypregn-4-ene-3,20-dione)		9.03
$C_{21}H_{30}O_5$	Cortisol		9.03

$C_{21}H_{32}O_3$ 4-Pregnene-20β,21-diol-3-one(20β, 21-Dihydroxypregn-4-ene-3-one)   9.02

C_{22}

$C_{22}H_{18}$	9,10-Dimethyltriptycene	11.11
$C_{22}H_{18}O_2$	9,10-Dimethoxytriptycene	11.11
$C_{22}H_{18}O_4$	Dibenzyl phthalate	11.19
$C_{22}H_{22}O_4$	Dienestrol diacetate (3,4-Di-(p-acetoxyphenyl) hexa-2,4-diene)	9.11
$C_{22}H_{23}NO_7$	Narcotine	10.11
$C_{22}H_{30}O_2S$	4,4'-Thio-bis(6-t-butyl-o-cresol)	4.09
$C_{22}H_{30}O_2S_2$	Santonox R (4,4'-Dithio-bis-(6-t-butyl-m-cresol)	4.10
$C_{22}H_{30}O_5$	6α-Methylprednisolone (6α-Methyl-11β,17α, 21-trihydroxypregna-1,4-dien-3-one)	9.09
$C_{22}H_{32}O_2$	Vitamin A acetate	8.16 8.19

C_{23}

$C_{23}H_{21}NO_4$	1-(3,4-Dibenzyloxyphenyl)-2-nitro-trans-prop-1-ene	11.07
$C_{23}H_{26}N_2O_4$	Brucine	10.05
$C_{23}H_{29}FO_6$	9α-Fluoroprednisolone acetate (17α-Acetoxy-9α-fluoro-11β,21-dihydroxypregna-1, 4-dien-3-one)	9.09
$C_{23}H_{30}O_6$	Cortisone 21-acetate	9.02
	Prednisolone acetate (17α-Acetoxy-11β, 21-dihydroxypregna-1,4-dien-3,20-dione)	9.10
$C_{23}H_{32}O_2$	4,4'-Methylenebis-(6-t-butyl-o-cresol)	4.09

C_{24}

$C_{24}H_{12}$	Coronene	1.01
$C_{24}H_{18}$	o-Quaterphenyl	1.02
	m-Quaterphenyl	1.02 1.03
$C_{24}H_{28}O_4$	Diethylstilbestrol dipropionate (3, 4-Di-(p-propionyloxyphenyl)hex-3-ene)	9.11
$C_{24}H_{32}O_6$	6α-Methylprednisolone acetate (17α-Acetoxy-6α-methyl-11β,21-dihydroxypregna-1, 4-dien-3-one)	9.09

$C_{24}H_{32}O_8$	Oestradiol-17α-glucosiduronic acid	9.12
$C_{24}H_{38}O_4$	Di(iso-octyl) phthalate	11.18

C_{25}

$C_{25}H_{32}N_4O_5$	Dehydro*epi*androsterone, dinitrophenylhydrazone derivative	9.05
$C_{25}H_{32}O_4$	Melengestrol acetate (17α-Acetoxy-6-methyl-16-methylenepregna-1,4-dien-3-one)	9.09
$C_{25}H_{34}N_4O_5$	5α-Androstane-3β-ol-17-one, dinitrophenylhydrazone derivative	9.05

C_{26}

$C_{26}H_{42}O_4$	Dinonyl phthalate	11.19

C_{27}

$C_{27}H_{33}N_9O_{15}P_2$	Flavin adenine dinucleotide	7.08
$C_{27}H_{44}O$	Vitamin D_3 (Cholecalciferol)	8.16
$C_{27}H_{44}O_6$	β-Ecdysone((22R)-2β,3β,14α,22,25-Pentahydroxy-5β-cholest-7-en-6-one)	9.04
	Ponasterone A((20R,22R)-2β,3β,14α,20,27-Pentahydroxy-5β-cholest-7-en-6-one)	9.04
	Ponasterone B((20R,22S)-2α,3α,14α,20,27-Pentahydroxy-5β-cholest-7-en-6-one)	9.04

C_{28}

$C_{28}H_{38}N_2O_4$	Emetine		10.09
$C_{28}H_{44}O$	Vitamin D_2 (Calciferol)	8.07	8.13
$C_{28}H_{46}O_4$	Di(isodecyl) phthalate		11.18

C_{29}

$C_{29}H_{40}N_2O_4$	Cephaeline		10.09
$C_{29}H_{50}O_2$	Vitamin E (α-Tocopherol)	8.07	8.13
$C_{29}H_{44}O_2$	4,4′-Methylenebis(2,6-di-t-butylphenol)		4.09

C_{30}

$C_{30}H_{22}$	*m*-Quinquephenyl	1.03

$C_{30}H_{54}$	Didodecylbenzene	1.10

C_{31}

$C_{31}H_{52}O_2$	Vitamin E acetate (α-Tocopherol acetate)	8.16

C_{35}

$C_{35}H_{62}O_3$	Irganox 1076 (Octadecyl 3,5-di-t-butyl-4-hydroxyhydrocinnamate)	4.10

C_{36}

$C_{36}H_{26}$	*m*-Sexiphenyl	1.03
$C_{36}H_{60}O_2$	Vitamin A palmitate	8.13

C_{38}

$C_{38}H_{47}NO_{13}$	3-Formyl Rifamycin SV	8.01

C_{40}

$C_{40}H_{56}$	Carotenes		10.01
	α-Carotene		10.02
	β-Carotene	8.16	10.02
$C_{40}H_{56}O_2$	Lutein		10.01
$C_{40}H_{56}O_4$	Neoxanthin		10.01
	Violaxanthin		10.01
$C_{40}H_{70}O_{14}$	Digitoxin		9.14

C_{41}

$C_{41}H_{56}N_4O_{11}$	Desacetyl Rifampin	8.01
$C_{41}H_{64}O_{14}$	Digoxin	9.14

C_{43}

$C_{43}H_{56}N_4O_{12}$	Rifampin quinone	8.01
$C_{43}H_{58}N_4O_{12}$	Rifampin	8.01

C_{51}

$C_{51}H_{72}O_3$	1,3,5-Tris(3,5-di-t-butyl-4-hydroxyphenyl)-2,4,6-trimethylbenzene	4.09

C_{63}

$C_{63}H_{88}CoN_{14}O_{14}P$	Vitamin B$_{12}$ (Cyanocobalamin)	8.07

Ga

Ga	Gallium	11.01

In

In	Indium	11.01

Na

$NaNO_2$	Sodium nitrite	11.02
$NaNO_3$	Sodium nitrate	11.02

O

O_2	Oxygen	6.04

Tl

Tl	Thallium	11.01

(The numbers refer to the chromatograms. Anions of carboxylic acids and sulfonic acids are listed under the parent acid.)

A

220

A

Adenosine-**5**′-monophosphoric acid $C_{10}H_{14}N_5O_7P$ 7.01   7.03   7.07   7.08

Adenosine-5′-triphosphoric acid $C_{10}H_{16}N_5O_{13}P_3$ ·7.01

Aflatoxin B_1 $C_{17}H_{12}O_6$ 10.04

Aflatoxin B_2 $C_{17}H_{14}O_6$ 10.04

Aflatoxin G_1 $C_{17}H_{12}O_7$ 10.04

Aflatoxin G_2 $C_{17}H_{14}O_7$ 10.04

Aldosterone $C_{21}H_{28}O_5$ 9.02

Aldrin $C_{12}H_8Cl_6$ 6.10   6.14

4-Aminoazobenzene $C_{12}H_{11}N_3$ 3.12

2-Aminobenzimidazole $C_7H_7N_3$ 11.05

2-Amino-5,6-dimethylbenzimidazole $C_9H_{11}N_3$ 11.05

Amobarbital $C_{11}H_{18}N_2O_3$ 8.06

Δ-1,4-Androstadiene-17β-ol-3-one $C_{19}H_{26}O_2$ 9.06

5α-Androstane-3β-ol-17-one dinitrophenylhydrazone $C_{25}H_{34}N_4O_5$ 9.05

4-Androstene-3,17-dione $C_{19}H_{26}O_2$ 9.05   9.06

Androst-4-ene-3,17-dione see 4-Androstene-3,17-dione   9.05   9.06

4-Androstene-11β-ol-3,17-dione $C_{19}H_{26}O_3$ 9.06

4-Androstene-3,11,17-trione $C_{19}H_{24}O_3$ 9.06

Androst-4-ene-3,11,17-trione, see 4-Androstene-3,11,17-trione   9.06

Androsterone $C_{19}H_{30}O_2$ 9.07

Aniline C_6H_7N 3.03   3.05

p-Anilinesulfonic acid $C_6H_7NO_3S$ 5.06

Anisole C_7H_8O 11.24

Anthracene $C_{14}H_{10}$ 1.06   1.07   1.09

Anthraquinone $C_{14}H_8O_2$ 11.15

Ascorbic acid, see Vitamin C   8.07

Aspirin $C_9H_8O_4$ 8.10   8.11

Azobenzene $C_{12}H_{10}N_2$ 3.01   3.12

1-Azepinyl-3-benzenesulfonylurea, see Tolazamide   8.02

Azulene $C_{10}H_8$ 1.02

B

B

C

C

o-Chlorophenol C_6H_5ClO		4.05
p-Chlorophenol C_6H_5ClO		4.05
2-Chlorophenothiazine $C_{12}H_8ClNS$		11.09
3-Chlorophenothiazine $C_{12}H_8ClNS$		11.09
p-Chlorophenyl o-nitrophenyl sulfide $C_{12}H_8ClNO_2S$		11.09
1-p-Chlorophenyl-3,3-dimethylurea, see Monuron		6.01
7-Chloro-1,2,3,4-tetrahydro-2-methyl-4-oxo-3-o-tolyl-6-quinazolinesulfonamide, see Metolazone		8.20
Chlorpropamide $C_{10}H_{13}ClN_2O_3S$		8.02
Cholecalciferol, see Vitamin D_3		8.16
Chrysene $C_{18}H_{12}$	1.05	1.06
Cinnamyl alcohol $C_9H_{10}O$		11.17
Citric acid $C_6H_8O_4$		2.07
Clopidol $C_7H_7Cl_2NO$		8.08
Codeine $C_{18}H_{21}NO_3$		10.11
Codeine phosphate $C_{18}H_{24}NO_4P$		8.11
Coronene $C_{24}H_{12}$		1.01
Corticosterone $C_{21}H_{30}O_4$		9.02
Cortisol $C_{21}H_{30}O_5$		9.03
Cortisone $C_{21}H_{28}O_5$		9.02
Cortisone 21-acetate $C_{23}H_{30}O_6$		9.02
Coumarin $C_9H_6O_2$		10.07
o-Cresol C_7H_8O		4.04
m-Cresol C_7H_8O	4.02	4.04
p-Cresol C_7H_8O		4.02
Cryptopine $C_{21}H_{23}NO_5$		10.11
Cyanocobalamin, see Vitamin B_{12}		8.07
3',5'-Cyclic adenosine monophosphate $C_{10}H_{12}N_5O_6P$	7.03	7.07
9-(4-c-Cyclopropyl-α-D-ribo-tetrafuranosyl)adenine $C_{12}H_{15}N_5O_3$		7.06
9-(4-c-Cyclopropyl-β-D-ribo-tetrafuranosyl)adenine $C_{12}H_{15}N_5O_3$		7.06
Cytosine $C_4H_5N_3O$		7.05
Cytidine-5'-diphosphoric acid $C_9H_{15}N_3O_{11}P_2$		7.01

C

Cytidine-2'-monophosphoric acid $C_9H_{14}N_3O_8P$	7.02	7.04
Cytidine-3'-monophosphoric acid $C_9H_{14}N_3O_8P$	7.02	7.04
Cytidine-5'-monophosphoric acid $C_9H_{14}N_3O_8P$	7.01	7.07
Cytidine-5'-triphosphoric acid $C_9H_{16}N_3O_{14}P_3$		7.01

D

p,p'-DDD $C_{14}H_{10}Cl_4$	6.10	6.14
o,p'-DDT $C_{14}H_9Cl_5$		6.14
p,p'-DDT $C_{14}H_9Cl_5$	6.10	6.14
Decalin $C_{10}H_8$		1.02
Dehydroacetic acid $C_8H_8O_4$		2.09
11-Dehydroepiandrosterone dinitrophenylhydrazone $C_{25}H_{32}N_4O_5$		9.05
11-Dehydrocorticosterone $C_{21}H_{28}O_4$		9.02
Deoxycorticosterone $C_{21}H_{30}O_3$		9.02
11-Deoxycortisol $C_{21}H_{30}O_4$		9.02
Desacetyl Rifampin $C_{41}H_{56}N_4O_{11}$		8.01
Dexedrine $C_9H_{15}NO_4S$		8.14
4,4'-Diacetamidodiphenyl sulfone $C_{16}H_{16}N_2O_4S$		8.04
3,4-Di-(p-acetoxyphenyl)hexa-2,4-diene, see Dienestrol diacetate		9.11
Di-(o-aminophenyl)methane $C_{13}H_{14}N_2$		3.02
Di-(m-aminophenyl)methane $C_{13}H_{14}N_2$		3.02
Di-(p-aminophenyl)methane $C_{13}H_{14}N_2$		3.02
4,4'-Diaminodiphenyl sulfone $C_{12}H_{12}N_2O_2S$		8.04
Diazepam $C_{16}H_{13}ClN_2O$		8.09
Diazepam (N-demethylated metabolite) $C_{15}H_{11}ClN_2O$		8.09
Diazepam (hydroxylated metabolite) $C_{16}H_{13}ClN_2O_2$		8.09
13H-Dibenzo[a,g]fluorene $C_{21}H_{14}$		1.01
Dibenzyl phthalate $C_{22}H_{18}O_4$		11.19
3,4-Dibenzyloxybenzaldehyde $C_{21}H_{18}O_3$		11.07
1-(3,4-Dibenzyloxyphenyl)-1-nitro-trans-prop-1-ene $C_{23}H_{21}NO_4$		11.07

D

2,6-Dimethylaniline $C_8H_{11}N$ 3.03 3.04 3.05

1,4-Dimethylanthraquinone $C_{16}H_{12}O_2$ 11.15

N,N-Dimethylbenzenesulfonamide $C_8H_{11}NO_2S$ 8.12

2,5-Dimethylbenzenesulfonic acid $C_8H_{10}O_3S$ 5.02

5,6-Dimethylbenzimidazole $C_9H_{10}N_2$ 11.05

5,6-Dimethyl-2-benzimidazolemethanol $C_{10}H_{12}N_2O$ 11.05

1,1-Dimethyl-3-[*m*-(t-butylcarbamoyloxy)phenyl]urea, see Tandex[R]   6.11

(2,3-Dimethylnaphthalene)tricarbonychromium $C_{15}H_{12}CrO_3$ 11.03

4,6-Dimethylorthanilic acid $C_8H_{11}NO_3S$ 5.03

1,1-Dimethyl-3-phenylurea, see Fenuron   6.01
2,6-Dimethylpyridin-4-ol C_7H_9NO 2.09

N,N-Dimethyl-*p*-toluenesulfonamide $C_9H_{13}NO_2S$ 8.12

9,10-Dimethyltriptycene $C_{22}H_{18}$ 11.11

2,4'-Dinitrobiphenyl $C_{12}H_8N_2O_4$ 11.13

2,6-Dinitrotoluene $C_7H_6N_2O_4$ 11.13

Dinonyl phthalate $C_{26}H_{42}O_4$ 11.19

Diphenylamine $C_{12}H_{11}N$ 3.06

Diphenylhydantoin $C_{15}H_{12}N_2O_2$ 8.03

2,5-Diphenyl-6a-selenathiophthen $C_{17}H_{12}S_2Se$ 11.08

2,5-Diphenyl-6a-thiathiophthen $C_{17}H_{12}S_3$ 11.08

3,4-Di-(*p*-propionyloxyphenyl)hex-3-ene, see
    Diethylstilbestrol dipropionate   9.11
4,4'-Dithio-bis-(6-t-butyl-*m*-cresol), see Santonox R   4.10
Diuron $C_9H_{10}Cl_2N_2O$ 6.01

Dodecanoic acid $C_{12}H_{24}O_2$ 2.05

Dodecylbenzene $C_{18}H_{30}$ 1.10

Dyfonate[R] $C_{10}H_{15}O_2PS_2$ 6.06

E

EPN[R] $C_{14}H_{15}NO_4PS$ 6.14

β-Ecdysone $C_{27}H_{44}O_6$ 9.04

Emetine $C_{28}H_{38}N_2O_4$ 10.09

Endrin $C_{12}H_8Cl_6O$ 6.10 6.14

Ephedrine $C_{10}H_{15}NO$ 8.14 8.15

Equilinen $C_{18}H_{18}O_2$ 9.12

7-Ethoxy-4-methylcoumarin $C_{12}H_{12}O_3$ 10.07

O-Ethyl (O-p-nitrophenyl)phenylphosphorothionate, see EPN[R] 6.14

N-Ethylaniline $C_8H_{11}N$ 3.06

2-Ethylanthraquinone $C_{16}H_{12}O_2$ 11.15

Ethylene glycol $C_2H_6O_2$ 11.20

Ethylhexyl 2,4-dichlorophenoxyacetate, see 2,4-Dichlorophenoxyacetic
    acid, ethylhexyl ester 6.07

5-Ethyl-5-(3-hydroxy-3-methylbutyl)barbituric acid, see
    Hydroxyamobarbital 8.06

5-Ethyl-5-(p-hydroxyphenyl)barbituric acid, see
    Hydroxyphenobarbital 8.06

5-Ethyl-5-isopentylbarbituric acid, see Amobarbital 8.06

5-Ethyl-5-phenylbarbituric acid, see Phenobarbital 8.06 8.11 8.15

F

FD&C Yellow 5 $C_{16}H_{12}N_4O_9S_2$ 5.06

Fenuron $C_9H_{12}N_2O$ 6.01

Flavin adenine dinucleotide $C_{27}H_{33}N_9O_{15}P_2$ 7.08

Flavin mononucleotide $C_{17}H_{21}N_4O_9P$ 7.08

Fluoranthene $C_{16}H_{10}$ 1.06 11.24

Fluorene $C_{13}H_{10}$ 1.05 1.07

9α-Fluoroprednisolone acetate $C_{23}H_{29}FO_6$ 9.09

Folpet $C_9H_4Cl_3NO_2S$ 6.02

3-Formyl Rifamycin SV $C_{38}H_{47}NO_{13}$ 8.01

Fumaric acid $C_4H_4O_4$ 2.01 2.08

G

G salt $C_{10}H_8O_7S_2$ 5.04

Gallium Ga 11.01

Glutaric acid $C_5H_8O_4$ 2.08

Guanine $C_5H_5N_5O$ 7.05

Guanosine-5'-diphosphoric acid $C_{10}H_{15}N_5O_{11}P_2$ 7.01

Guanosine-2'-monophosphoric acid $C_{10}H_{14}N_5O_8P$ 7.02

Guanosine-3'-monophosphoric acid $C_{10}H_{14}N_5O_8P$ 7.02 7.04

Guanosine-5'-monophosphoric acid $C_{10}H_{14}N_5O_8P$ 7.01 7.07

Guanosine-5'-triphosphoric acid $C_{10}H_{16}N_5O_{14}P_3$ 7.01

H

Heptachlor $C_{10}H_5Cl_7$ 6.14

1,4,5,6,7,8,8-Heptachloro-3a,4,7,7a-tetrahydro-4,7-methanoindene,
    see Heptachlor   6.14

Hexachlorobenzene C_6Cl_6 11.14

γ-1,2,3,4,5,6-Hexachlorocyclohexane, see Lindane$^{(R)}$ 6.10 6.14

1,2,3,4,10,10-Hexachloro-6,7-epoxy-1,4,4a,5,6,7,8,8a-octahydro-1,
    4-*endo-endo*-5,8-dimethanonaphthalene, see Endrin   6.10 6.14

1,2,3,4,10,10-Hexachloro-1,4,4a,5,8,8a-hexahydro-1,
    4-*endo-exo*-5,8-dimethanonaphthalene, see Aldrin   6.10 6.14

Hexanoic acid $C_6H_{12}O_2$ 2.05

Hydroquinone $C_6H_6O_2$ 4.06 4.07

p-Hydroxyacetanilide, see Paracetamol   8.10

Hydroxyamobarbital $C_{11}H_{18}N_2O_4$ 8.06

17β-Hydroxyandrosta-1,4-dien-3-one, see Δ^1-Testosterone   9.05

17β-Hydroxyandrosta-1,4-dien-3-one, see Δ-1,
    4-Androstadiene-17β-ol-3-one   9.06

3β-Hydroxy-5α-androstan-17-one 2,4-dinitrophenylhydrazone,
    see 5α-Androstane-3β-ol-17-one dinitrophenylhydrazone   9.05

11β-Hydroxyandrost-4-ene-3,17-dione, see 4-Androstene-11β-ol-3,
    17-dione   9.06

17β-Hydroxyandrost-4-en-3-one, see Testosterone   9.05 9.06 9.07

3β-Hydroxyandrost-5-en-17-one 2,4-dinitrophenylhydrazone,
    see Dehydro*epi*androsterone dinitrophenylhydrazone   9.05

p-Hydroxybenzenesulfonic acid $C_6H_6O_4S$ 5.07

4'-Hydroxybiphenyl-4-sulfonic acid $C_{12}H_{10}O_4S$ 5.01

3-Hydroxybutyric acid $C_4H_8O_3$ 2.08

6β-Hydroxycortisone $C_{21}H_{28}O_6$ 9.02 9.03

4-Hydroxy-2,6-dimethylpyridine-3-carboxylic acid $C_8H_9NO_3$ 2.09

2-Hydroxy-4-methoxybenzophenone $C_{14}H_{12}O_3$ 4.09

7-Hydroxy-6-methoxycoumarin $C_{10}H_8O_4$ — 10.07

3-Hydroxy-2,7-naphthalenedisulfonic acid, see R salt — 5.04

7-Hydroxy-1,3-naphthalenedisulfonic acid, see G salt — 5.04

6-Hydroxy-2-naphthalenesulfonic acid, see Schaeffer's salt — 5.04

3-Hydroxyoestra-1,3,5(10)-triene-16,17-dione, see 16-Keto-oestrone — 9.12

17β-Hydroxyoestr-4-en-3-one, see 19-Nortestosterone — 9.06 9.07

Hydroxyphenobarbital $C_{12}H_{12}N_2O_4$ — 8.06

5-p-Hydroxyphenyl-5-phenylhydantoin $C_{15}H_{12}N_2O_3$ — 8.03

17α-Hydroxypregn-4-ene-3,11,20-trione, see 4-Pregnene-17α-ol-3, 11,20-trione — 9.03

21-Hydroxypregn-4-ene-3,11,20-trione-see 11-Dehydrocorticosterone — 9.02

Hypoxanthine $C_5H_4N_4O$ — 7.05

Imidan $C_{11}H_{12}NO_4PS_2$ — 6.02

Indane C_9H_{10} — 1.05

Indene C_9H_8 — 1.04 1.07

Indium In — 11.01

Inosine-5'-monophosphoric acid $C_{10}H_{13}N_4O_8P$ — 7.03 7.07

Irganox 1076 $C_{35}H_{62}O_3$ — 4.10

Isobutyl 2,4-dichlorophenoxyacetate, see 2,4-Dichlorophenoxyacetic acid, isobutyl ester — 6.07

Isocitric acid $C_6H_8O_7$ — 2.07

Isonicotinamide $C_6H_6N_2O$ — 3.13

Isonicotinonitrile $C_6H_4N_2$ — 3.14

Isophthalic acid $C_8H_6O_4$ — 2.04

Isopropyl 2,4-dichlorophenoxyacetate, see 2,4-Dichlorophenoxyacetic acid, isopropyl ester — 6.07

4,4'-Isopropylidenediphenol, see Bisphenol A — 4.01

Isoquinoline C_9H_7N — 3.01 3.09

K

Ketohexobarbital $C_{12}H_{14}N_2O_4$ — 8.06

16-Keto-oestradiol $C_{18}H_{22}O_3$ — 9.12

16-Keto-oestrone $C_{18}H_{20}O_3$ — 9.12

2-Methyl-1,4-naphthoquinone, see Vitamin K_3 8.07

(+)-α-Methylphenethylammonium hydrogen sulfate,
    see Dexedrine   8.14

6α-Methylprednisolone $C_{22}H_{30}O_5$ 9.09

6α-Methylprednisolone acetate $C_{24}H_{32}O_6$ 9.09

6-Methylpyridine-2,4-diol $C_6H_7NO_2$ 2.09

N-Methyl-p-toluenesulfonamide $C_8H_{11}NO_2S$ 8.12 11.23

6α-Methyl-11β,17α,21-trihydroxypregna-1,4-dien-3-one,
    see 6α-Methylprednisolone   9.09

Metolazone $C_{16}H_{16}ClN_3O_3S$ 8.20

Monuron $C_9H_{11}ClN_2O$ 6.01

Morphine $C_{17}H_{19}NO_3$ 10.11

N

Naphthalene $C_{10}H_8$ 1.02   1.04   1.05   1.06
     1.07   1.09 11.24

Naphthalene-1,5-diol $C_{10}H_8O_2$ 4.07

Naphthalene-2-sulfonic acid $C_{10}H_8O_3S$ 5.02

1-Naphthol $C_{10}H_8O$ 4.06   6.08

2-Naphthol $C_{10}H_8O$ 4.06

1-Naphthylamine $C_{10}H_9N$ 3.01

2-Naphthylamine $C_{10}H_9N$ 3.01 11.21

1-Naphthyl-N-methylcarbamate, see Carbaryl   6.08

Narcotine $C_{22}H_{23}NO_7$ 10.11

4-Neopentyl-1,2-dithiole-3-thione $C_8H_{12}S_3$ 11.06

Neoxanthin $C_{40}H_{56}O_4$ 10.01

Nicotinamide $C_6H_6N_2O$ 3.13   3.14

Nicotinonitrile $C_6H_4N_2$ 3.13   3.14

o-Nitroaniline $C_6H_6N_2O_2$ 3.08

m-Nitroaniline $C_6H_6N_2O_2$ 3.08

p-Nitroaniline $C_6H_6N_2O_2$ 3.08

p-Nitrobenzenesulfonamide $C_6H_6N_2O_4S$ 11.23

3-Nitrofluoranthene $C_{16}H_9NO_2$ 11.13

3-Nitro-4-methoxybenzamide $C_8H_8N_2O_4$ 11.23

O

P

U

Uracil $C_4H_4N_2O_2$			7.05
Uridine-5'-diphosphoric acid $C_9H_{14}N_2O_{12}P_2$			7.01
Uridine-2'-monophosphoric acid $C_9H_{13}N_2O_9P$		7.02	7.04
Uridine-3'-monophosphoric acid $C_9H_{13}N_2O_9P$		7.02	7.04
Uridine-5'-monophosphoric acid $C_9H_{13}N_2O_9P$		7.01	7.07
Uridine-5'-triphosphoric acid $C_9H_{15}N_2O_{15}P_3$			7.01

V

Violaxanthin $C_{40}H_{56}O_4$			10.01
Vitamin A $C_{20}H_{30}O$	8.07	8.13	8.19
Vitamin A acetate $C_{22}H_{32}O_2$		8.16	8.19
Vitamin A palmitate $C_{36}H_{60}O_2$			8.13
Vitamin B_2, see Riboflavin		7.08	8.07
Vitamin B_{12} $C_{63}H_{88}CoN_{14}O_{14}P$			8.07
Vitamin C $C_6H_8O_6$			8.07
Vitamin D_2 $C_{28}H_{44}O$		8.07	8.13
Vitamin D_3 $C_{27}H_{44}O$			8.16
Vitamin E $C_{29}H_{50}O_2$		8.07	8.13
Vitamin E acetate $C_{31}H_{52}O_2$			8.16
Vitamin K_3 $C_{11}H_8O_2$			8.07

X

m-Xylene C_8H_{10}		3.03
m-Xylenetricarbonylchromium $C_{11}H_{10}CrO_3$		11.04
2,3-Xylenol $C_8H_{10}O$		4.02
2,4-Xylenol $C_8H_{10}O$		4.04
2,6-Xylenol $C_8H_{10}O$	4.02	4.03
3,4-Xylenol $C_8H_{10}O$		4.02
3,5-Xylenol $C_8H_{10}O$		4.02